THE
HAIR
COLOR
MIX
BOOK

INTRODUCTION

THE INNER HAIR COLORIST

I have been so honored to work with the crème de la crème in my field. I have been stimulated and exposed to so much talent over the years, which has inspired me to invent original techniques and formulas. I want to give back to both aspiring and dedicated at-home hair colorists by revealing some of what I have been privileged to learn over twenty-one-plus years of salon work.

The talented actresses and actors that I have worked with have also stretched my abilities. Through the years I have had the privilege to work on many films, coloring the hair of many amazing men and women, including Reese Witherspoon, Heidi Klum, Mary-Kate and Ashley Olsen, Raquel Welch, Brad Pitt, and Johnny Depp. Along the way, lucky stylists encounter that rare client who helps them ascend to the next level of creating great color. For me, Drew Barrymore is that client—she has always pushed me to a higher plane of visual perfection!

Nowadays, my clients fly in from New York City, London, and all over the world to see me, and I get to do long-distance house calls. I've gone from Beverly Hills to Africa. I have sent recipes all over the world as well, including Madonna's homes in New York and London. It still utterly amazes me that hair coloring has enabled me to travel all over the world. I used to guard my color formulas as if they were twenty-four-karat gold. I would take them home every night from the salon. As I FedExed recipes to clients across the globe, more frequently, I began thinking about my policy of keeping my recipes a secret. I remembered that when I was in my teens, I wanted to emulate the models and actresses in magazines. I didn't exactly have the budget to do so, but I did have the creativity, desire, and drive. I think there are many people like that. They want to do it themselves.

I want to share my favorite recipes and techniques because, at heart, I am just like you. I have the same spirit of invention and optimism that all home colorists share. Even though I'm based in Hollywood and work with many celebrities, movie stars, and film-industry executives, I think of myself as the ultimate at-home colorist. I love to experiment, try new things, and have fun. I still get excited when I discover a new technique or when the result of a new recipe is extra special.

Hair coloring, after all, is much like cooking, painting, or any other creative endeavor. First you learn the rules, and then you learn how to break them and find your own style. Actually, I think my very earliest artistic endeavors were a sign of things to come. I started painting when I was three or four years old. My wonderful stepfather, Jon, was an architect, and he and my mother, Connie, had a keen interest in art and creativity. After school, they would give me three dollops of primary color paint to fool around with. Both of them painted and he sculpted, so I felt intimately involved in the artistic process from a very early age. Without Jon and my mother's encouragement, I would not be who I am today.

It wasn't long before I put down my finger paints and turned my attention to hair color. Early experiments included adding Sun-In, the classic spray-on lightener, to my hair and styling my mother's and sister's hair and doing their makeup. However, hair color didn't become a true passion of mine until I was on summer break while studying as a nursing student and special education major at Centralia Community College in Washington State.

My aunt Peggy Olsen owned a salon called the Curling Iron in Upland, California. In fact, she won many "beehive trophies" for the elaborate hairstyles she constructed! During this particular summer break, I went to Upland and worked as a receptionist in the salon. As I quietly kept the books and scheduled clients, I would look around and think, *Wow, a salon allows you to have fun, be creative, make money, and meet interesting people!* Redken representatives came in to the Curling Iron and would conduct detailed color classes. I started sitting in on them. Redken was a major hair company (now owned by L'Oréal), and its approach to hair coloring was very scientific and clinical. Their seminars opened my eyes even further about the possibilities for hair-color work. I think it was at that point that I thought to myself, *Okay, that's it, I'm going to hair school.*

I didn't want to spend years studying in a classroom, so I found a job and enrolled in night school. After graduating from Citrus Community College in Glendora, California, I found my first professional employment through an ad in the *Los Angeles Times* that read, "European Salon seeks assistant." That sounded pretty glamorous to me, so I applied. Well, it turned out to be a tiny mom-and-pop beauty parlor. The only European thing about it was the stunning Russian-Jewish women who worked there and the owner's wife, who was from England and was a great manicurist. Nevertheless, that's where, over the next two years, I began to hone my hair-coloring skills.

I commuted from LaVerne to Los Angeles for an entire year—a one-hour trip each way. But I didn't care: I wanted to learn the trade I had come to love. When I finally moved to Beverly Hills with a girlfriend, it was easier for me to go out in the evenings and socialize, and that changed my life. Tramp was a fabulous 1980s members only club where lots of rockers, such as Prince, the members of Mötley Crüe, and Rod Stewart, went. I met Laurent Dufourg and his wife, Fabienne (now of Privé fame), in the club's dining section. They had partnered with master hair stylist José Eber and were opening a new salon. Laurent suggested that I try out for a spot.

That was an exciting moment, so I called the salon and made appointments to bring models and do hair color that would show my work. For some time after those initial meetings, nothing happened except that I kept running into Laurent and Fabienne at many different events. Finally, after a year, I found out that José was ready to open the salon. José's secretary called me and asked me to come in for an interview. After that call, I thought it was a done deal. Then, I walked into the salon three days later and found a line of 150 people waiting to compete for the twenty available spots! In the end, I was hired despite the competition. My trademark was to pamper clients with luxury in hopes of building up a clientele. If you treat people well, they usually come back.

At that job I also learned much about corrective color, which I had never done before. I was with José Eber for two and a half years, where I was inspired by the amazing Corey Powell, and then relocated to Christophe in Beverly Hills, where I began working with stylist extraordinaire Olivier Leroy (this was right before the famous Clinton haircut). I was there for eight and a half years. Afterward I moved to Jonathan Hair Salon, where I met and began to work with überstylists Sally Hershberger and Cervando Maldonado, then to John Frieda, and now I am at Neil George Salon in Beverly Hills. I feel privileged and thankful that my clients have stayed true to me.

There's another reason that I want to share my hair-color knowledge. Some people may think working with hair is pretty trivial because it isn't brain surgery or calculus. However, I know that if people feel good about themselves, they just may change the world. Having the power to improve oneself or the way one looks is just the beginning of a journey toward confidence and well-being. Besides, all of us know how confident we feel when our hair looks and feels great!

I believe everyone should have access to beautiful hair, no matter what a person's financial or residential limitations are—whether they simply can't afford the high price of "couture color" or don't live near a city or town that offers highly trained colorists. That's what the recipes and salon secrets in this book do. You can do it—I believe in you. All you need is the information in this book, a few local stores, and/or Internet access. I hope on another level that this book inspires you to go after your own dreams. If you are sitting in a small town somewhere, like I was, wishing you could run a hair salon, work on a movie set, fly into space, become a doctor, teach agriculture, be a great mom, or whatever, I want you to know you really *can* do it if you put your mind to it. My story is just one example that proves that. So let's start playing, mixing, and having fun. You can do it! But like mastering cooking or knitting, it takes practice. So step-by-step, we'll do it together!

I look forward to seeing and hearing about your amazing results!

Take care and be well,

Lorri Goddard-Clark

PART I

The Basics

1.

REALITY CHECK

THINKING
OUTSIDE THE BOX

The beautiful models on the cover of the boxes of at-home color dyes have amazing hair, don't they? Absolutely! Despite that fact, have you ever been disappointed with your hair color after using an off-the-shelf dye? Do you ask yourself, "Why doesn't my hair turn out like the model's mane?" There are three simple reasons for the difference between the promise (what's on the box) and the outcome (what's on your head).

1. First, and most important, the model's natural color is most likely not the same as yours, so your results will be different from what's shown.

2. Second, the hair color shown on the cover of the box is achieved with more than just one single-process tint.

3. And finally, if the photo itself is highly manipulated, it rarely corresponds with what is realistically possible.

Let's examine reason number one because it is the premise on which this book is based. Every hair color responds in a unique way to the formula in a box color. For example, you are dark blonde, and your girlfriend is dark brown. If you both use the same brand of light golden blonde, each of you will achieve a different result. Your hair will have a lighter effect than your friend's. So the first step in creating the beautiful color *you* want is to identify your natural color using a simplified version of a professional hair-color tab system. Only then can you choose the right single color or mix up a tint recipe that corresponds to your natural shade to achieve what I call a target color, to achieve a specific effect. I have chosen a series of target colors in this book, and a version of almost every one of them is available to you, no matter your ethnicity or age.

There are so many great at-home hair-color products on the shelves of your favorite drugstores. In fact, there has never been a better time to be an at-home hair colorist! However, in terms of reason two, at-home hair-color kits do not tell you one vital fact: It is often necessary to use more than one color to get the most beautiful, natural look. In some instances, a carefully chosen single color can work. However, you may need to choose two, three, or maybe more boxes to achieve your objective. I have colored the hair of Milla Jovovich, Charlize Theron, and Elaine Irwin Mellencamp for major print and television-advertising campaigns for at-home hair color. A day or two before the shoot, the executives bring a photograph of a particular color and say, "Lorri, give us this color." I can't achieve the color they ask for with the ingredients in the boxes their pictures will end up on. I have to give the model an all-over dye (or what I like to call a "global color"), lowlights, and highlights.

Take a look around. Does Mother Nature work with only one shade? Colors in nature are not flat. Many shades and tones come together to create the depth, sheen, and glimmer we see in flora and fauna, animal fur, and human eyes and hair. The more variation you have in your hair, the more natural it looks. Keeping it real keeps hair gorgeous.

Hair-color professionals have a large cabinet full of shades that enable us to create the richest hair colors. These bottles contain the butter and lemon that give blonde hair light and sparkle, cinnamon and coffee tones that give brunettes sultry depth, and the merlot and apricot that give redheads the passion for which they're famous. The hair colorist's spice rack and pantry have the ingredients that add many levels of flavor and warmth to hair. And that richness, in turn, punches up eye colors and makes skin tones vibrant. A box of color from the store, no matter how good it is, usually needs a little help.

Applying multiple shades reacts with the natural variations of your hair to show off its "grain" in the same way polishing a wood surface with Pledge brings out its beauty and richness: different textures, ribbons of light—depth, warmth, and coolness. It's all about the placement of light and dark. A single, inappropriate color takes away natural variations and leaves you with a one-dimensional look.

UNREALITY CHECK

How about reason three? The final factor at play is the perfection required by advertising that necessitates giving a false impression of reality. So even after all the professional color work is done on the model's hair, the resulting professionally lighted photograph is retouched, airbrushed, and digitally enhanced. You can't digitally alter your hair, of course (not yet, anyway). But the good news is that this book reveals recipes and application techniques so you too can turn off-the-rack color into your own personal couture creation!

THE BIG PAY OFF!

The information in this book saves you money and time. The initial investment for setting up your personal salon is a bit more than the cost of one box of drugstore hair color, but in the long run, the cost of creating colors at home saves you money compared to what it would cost you to produce the same color in a salon. And you may already have many of the required items in your kitchen pantry or bathroom cabinet (Chapter 2). As I am sure you know, the price of coloring or highlighting hair, whether you are in New York, Miami, Los Angeles, Las Vegas, Biloxi, or Kansas City, is high. Correcting color mistakes is even more expensive! My recipes take the guesswork out of at-home color and will reveal how you can make professional custom formulas at home—with products available at drugstores and supermarkets! I can't be with all of you in person, but with this book, I hope you feel I am alongside you, cheering you on and supplying you with information that until now has not been available to the general public.

The recipes are cost-effective and effect-effective! They allow you to achieve the best possible results at home. And the salon techniques I share make application as easy, accurate, and highly efficient as possible.

Think of the newfound freedom creating your own salon at home gives you: Touching up your regrowth in between salon appointments is a breeze when you can do it your-

MY GOAL FOR AT-HOME COLORISTS EVERYWHERE

I will help you

1. Identify your natural hair color so you can choose the right recipes for the result you want.

2. Pick the right boxes from the drugstore shelf.

3. Mix box tints and other ingredients for a "beyond the box" color that's uniquely yours.

4. Use common products like cotton balls, Q-tips, and olive oil to help products glide on smoothly and precisely—and stay put once they're on!

5. Cover grays effectively for a natural look.

6. Use special triple-barrier cream around your face to prevent stains—and remove stains when they inevitably happen. (Even after twenty-plus years, I constantly make a mess. The box doesn't tell you how to clean up!)

7. Enhance your color with the help of an Oral-B toothbrush, plastic wrap, a hair dryer, a homemade diffuser, and other common household items for even better color results.

self, and that trip to the hair colorist you needed to book before a wedding or an interview or date can be postponed. And if the kids have the flu, you no longer need to leave the house for a color refreshing. In short, you can make an appointment with yourself, which is convenient for you—and buy yourself some holdover time until you can get back to the salon again.

IDENTITY CHECK

As I said in the beginning, before you delve into the recipes and start mixing, it's so important to know what you've got to work with before you begin. In fact, it's *crucial* for a successful result. It's the first step in making an informed color decision.

The eleven-color chart that can be found on pages 2 and 3 of the photo insert is a simplified, easy-to-use version of a tab system colorists have used for forty years. Every professional classification system uses levels to identify hair color. I have selected these eleven tabs because I believe they represent the most common natural colors I have seen in more than twenty years of professionally assessing hair.

Using my color-matching system, find the tab that most closely resembles your natural shade. That's the color growing out of the top of your head. People commonly refer to the hair growing out of their head a few weeks after they have color applied as their "roots," but technically, roots are *under* the scalp and invisible. Regrowth is the term for the color we see nearest our scalp. If you can examine your hair in natural light, do so. The next best thing is to look at it under a GE Reveal lightbulb. It does an excellent job of duplicating daylight.

When in doubt, choose the lighter of the color tabs that closely match your natural shade. You can always reapply a darker shade to get the desired result if you have to. Remember, color is the same as flavor in cooking. It's easy to add color (flavor), but difficult to take it away once it's been added.

If it makes matching easier, grab your mom, sister, or friend and help one another identify your tabs. Getting your tab right is important because it is the key to selecting the right recipes, since each one in the book is identified by tab number. The most natural color results can be achieved using recipes within two target-color tabs, in either direction, of your tab. So, for example, if you are a Tab 4, you can use recipes marked for target-color Tabs 2, 3, 4, 5, and 5½. You will hit the target color by using the recipe for your specific tab—a darker or lighter version is reached by using tabs in either direction from your tab's recipe. Any recipe that is more than two tabs away from your tab may not look as natural and requires plenty of maintenance, including regular brow tinting, frequent root touch-ups, and even additional makeup application.

LORRI'S SIMPLE TAB SYSTEM

Please check out pages 2–3 of the photo insert to see these actual colors:

TAB 1—Black

TAB 2—Darkest ash brown

TAB 3—Dark brown

TAB 4—Medium ash brown

TAB 5—Medium brown

TAB 5½—Light ash brown

TAB 6—Light brown

TAB 6½—Light reddish brown

TAB 7—Dark ash blonde

TAB 7½—Dark blonde

TAB 8—Medium blonde

TAB 9—Up to 35 percent white

TAB 10—35 percent to 65 percent white

TAB 11—65 percent to 100 percent white

NOTE: For those whose natural hair color resembles Tabs 9–11, I often suggest you revert to the recipe for your pre-gray color tab, or the color you were before you went gray. So you should identify two tabs—your percentage of gray tab (9–11) and your natural color tab (1–8), so you can use gray tab-recipe modifications and identify your natural tab recipe to complete your color look.

Once you have found your tab, the next step is to check out your eye color. Skin tone has a tendency to change with the seasons. Eye color is a more reliable way of judging, along with your natural hair color, the right tones (warm or cool) for highlights and lowlights.

Mirroring eye color in hair highlights and lowlights also adds incredibly natural looking dimension and depth to your hair. So don't skip this step, especially if you haven't colored your hair in a while or changed your salon or at-home hair color in several years. While our eye colors don't change drastically as we get older, they can become slightly lighter and warmer as we age, so it is worth giving them a look. Anyway, when was the last time you really looked into your own eyes? Doing so will be a pleasant surprise.

Using warm (yellow) tones vs. cool (blue) tones, depending on your eye color, makes all the difference. You are doing "makeup" for your whole face. Think of the way you apply your eye makeup: Highlighter and shine goes on the surface you want to pop, and darker colors combined with deeper colors help you create shadows and contrast. Playing up your hair color means showing off your eyes.

The very simple eye-color classification system on page 1 of the photo insert helps you use your eye color to determine appropriate highlight recipes.

- **DARK WITH WARM:** Examples are yellow tones, such as milk-chocolate brown and dark green; think gold or copper metal.

- **LIGHT WITH WARM:** Examples are yellow tones, such as tiger's-eye brown and green or blue with yellow flecks.

- **DARK WITH COOL:** Examples are blue tones, such as black, black-brown, dark chocolate, ivy green, and any dark gray or black flecks.

- **LIGHT WITH COOL:** Examples are blue tones, such as sky blue, lavender, sea foam, and blue-green.

If you are really having trouble determining your true eye color, go to your local fabric store and buy a yard of gold and a yard of silver or gray cloth. In natural daylight, drape one and then the other over your shoulder, and decide which shade looks more flattering on you. If gold looks best, you're a warm tone; if the silver or gray flatters your face, you are a cool tone. The warmth and coolness of highlights are indicated in the recipes. For example, I am one of the many people who have warm green eyes and ruddy skin. Warm colors and highlights look best on me because they show off my eyes. However, I have to be careful to modify my skin tone slightly by using a self-tanner. (Bain de Soleil is my favorite!)

MIRROR, MIRROR

Finally, you have to have what I call a "Looking-Glass Moment." Ask yourself, "What do I *really* like? What's my fantasy color? Am I happy with my hair color? Does it feel tired, faded, or washed out? Is my hair fried from too much color?" Using chemicals repetitively on hair and the scalp take their toll. Be warned: If you want to be much lighter than nature intended, then you must keep your hair a little shorter—or deal with frayed hair. In other words, if you want to keep your hair long, you shouldn't go dramatically lighter—unless you are a professional, and even then it can be tricky. Those celebrities you see with long blond hair that is not naturally light use extensions—all of them. Trust me.

Define what makes you feel hotter, sexier, prettier, sleeker, more stylish, fresher, younger, more in charge, ready for the office, ready for your man, ready for *you*. The color of our hair affects us psychologically. It is one of the ways that others form a perception of us. Ever notice the different impression Madonna makes when she's blonde versus when she's brunette? Do we prefer Jay Leno salt-and-pepper gray or dark-haired? Halle Berry conveys a girl-next-door look with soft brown hair with highlights; she's more sophisticated when she goes darker with deep lowlights. Nicole Kidman seems like a completely different person when she's blonde instead of red-haired. Brad Pitt is a California surfer when he's bleach-blond, and he's just plain handsome when his hair is softer and darker blond.

Tear into two or three of your favorite fashion or gossip magazines. If you don't look at those kinds of magazines on a regular basis, I suggest picking up a copy of *People* (*Teen*

People if you are under nineteen) and a comprehensive fashion magazine, such as *Vogue* or *Elle*. Every issue contains many current pictures of your favorite celebrities and trendsetters. Take a look—what styles appeal to you? If you have similar coloring to your favorite celebs, take a peek at how they color their hair. Check out their style. Do you love it? Rip out whatever appeals to you, and if a pattern emerges, it's a clue to what you really want in your own look. Why *not* fulfill a hair-color fantasy?

Think about your personality. Do you like being the center of attention? High-profile colors will steer eyes your way. Are you a little shy or prefer not to be perceived as a babelicious babe? Try a color that's closer to your own, and use highlights to frame and flatter your face in a subtle way.

Next, consider your lifestyle. Don't discount your personal daily grind! If you're running after little ones all day or have to file briefs first thing in the morning and then work until 9:00 PM, maybe you are better off with shades that cut your touch-up schedule some slack.

Be completely honest with yourself about how much time you can give to taking care of your hair and keeping it looking its best. Colors that are further in lightness or darkness

DRAMATIC COLOR, EXTRAORDINARY MEASURES

If you are determined to go three tabs in either direction of your natural color, you have to modify your makeup and skin tone. Are you willing to take the extra time to drag that lipstick around your lips, use a self-tanner, or wear extra foundation and blush? Do you have the time and the know-how to apply a little more eye shadow? If you want to be an Elizabeth Taylor brunette, for example, and it's not within your tab range, you have to enhance your makeup tones; otherwise, you will appear washed out, and your coloring will be left looking off balance. Making up every day takes time. Do you have that luxury?

from your natural color require more maintenance than colors that are nearer to your natural shade. Do you have time to look after your hair every two or three weeks? Or is a six- to eight-week schedule more realistic for you?

There are other questions you should ask: Does your significant other agree with your choice? Do you care about his or her opinion? Are you newly single or going through a big change in your life? Getting ready for a date with a guy who could potentially be the one? What about your career: Did you just get a new job? Are you in the process of looking for a new job? Are you trying an important court case? Will your boss perceive you as unstable or frivolous if you show up on Monday morning with a big color change—do you want that? Or do you work in an industry where that would be perceived as adventurous and you will be seen as a fearless go-getter? Are you a homemaker, artist, teacher, movie star, or lawyer? Do you want to be any one of those things?

SET THE PARAMETERS

Once you have identified your tab, evaluated your coloring, and assessed your lifestyle, it is time to choose a target color and get going! Now I want you to get real. Certain colors are achievable and appropriate at home. Other, more extreme desires need to be marched straight to the local licensed professional. Hair color is like many things in life: The more you do it, the better you get. Let's start out with a specific, doable goal, and after you get that right, we can progress to more advanced techniques. Even if you are a dark brunette and you've always wanted to be a blonde, you can create a lighter look and still look natural with recipes in this book simply by choosing two tabs in a lighter direction and adding face-framing highlights. The recipes will not give you a drastic change, but it is a flattering change that fulfills your desire to be lighter and brighter.

In general, slow and steady changes are usually best—and sometimes gradually changing hair color will not be consciously noticed by others. They will just think you look terrific, but they won't be sure why!

The recipes in this book are more about covering gray, highlighting, and darkening and lightening around your face. It's about creating plenty of little "moments" on the edges of your face that play up your eyes and brighten your face. In short, you'll look pretty. But that doesn't mean you can't get the impression of a more dramatic change. That's where the art of highlighting comes in. For example, if you are a Tab 1 and you've always dreamed of being Rita Hayworth red, you can get that *feeling* by using the appropriate version of red-hot highlights.

Once you find a vibe you like, we need to see if you can do it at home or have to go to the salon, or we need to adjust our picture in our minds and look for something a little different to start with. Don't forget, if you want to go light, it is better to start with subtle changes and add more highlights later. Again, if you want to go red, begin with a subtle shift; you can always go redder. If you want to obtain a deeper color than you already are, try for one to two shades darker by using a recipe for one or two tabs darker than you are.

SO MANY COLORS, SO LITTLE TIME

Read through the entire book first, familiarize yourself with the equipment (Chapter 2) and techniques (Chapter 3), and look at the celebrity photos in the color insert, which illustrate the target colors. Choose your favorite target color and read through the recipe for your tab before you begin. If there is no recipe for your tab, choose another similar, target color and find your tab. If the target colors you choose don't have recipes for your tab, that means those target colors are not for the at-home colorist to attempt, so hightail it to the salon! Radical changes are best left to professionals. Even then, don't expect miracles, but do expect to make regular visits to the colorist. (Can you afford the time and financial investment?) Next, set aside enough time

THERE ARE FOUR BASIC CUTS:

BASIC HAIRCUT No. 1

BASIC HAIRCUT No. 2

1. Very slight layers or one length, chin length, or longer without bangs

2. Very slight layers or one length, or longer with bangs

How you treat these four situations with high- and lowlights make or break your look. A #2 or #4 cut, for example, necessitates applying a lighter product on top of the bangs versus depth underneath. For cut #3, apply highlights around the hairline and outer edges.

I recommend getting your hair cut in advance rather than on the same day (or within one week) you are planning to do your color. That way you can see how it falls, and high- and lowlights (what I call "the sunshine") can be accurately placed.

BASIC HAIRCUT No. 3

3. Short to chin-length layered hair worn back or off the face

BASIC HAIRCUT No. 4

4. Quite short (chin length) with layers all around the face and forward to the face

to complete the color process carefully, gather materials, prepare your work area, and get started. Have fun!

I want all of you color-mix girls out there to slow down—you don't have to be the Road Runner. Make an appointment with yourself, just as you would with a salon. Carve out enough time to follow all the steps and complete your transformation. Read through the recipe and instructions for your chosen color and generally estimate the time it will take to complete—with the in-between shampooing and general preparation of your work area, that could be anywhere from two to four hours! Make a pleasant morning or afternoon out of it. Actually, if you can, it's preferable to do your color consultation in the morning or sometime during full daylight hours. Bring your mirror over to the window and look. Natural light gives you the best chance to assess your color accurately.

It is important to be playful, yet stay realistic, when deciding on a hair color. The first thing I like to do in the salon with a new client or a client wanting a big change is pull out the magazines. Or I'll encourage my clients to bring in photos of hair colors they aspire to and are inspired by.

One more thing: I do not work for any of the companies or brands I mention throughout this book or in the Resource section. I have tested a variety of products, and I name names because I have found so many brands to be high quality and reliable for the at-home colorist.

HOW TO USE THIS BOOK

1. Read through the entire book.

2. Find your tab.

3. Assess your eye color (key for the at-home color consultation!).

4. Make time for your at-home salon experience.

5. Gather your materials and ingredients.

6. Choose your recipe.

7. Prepare your work area.

8. Color your hair.

9. Get out there and stop traffic!

So grab those measuring cups and tools from your own at-home hair-color kit, then head to your local drugstore or supermarket, or check out some online resources, and let's get gorgeous!

2

PANTRY RAID

ESSENTIAL EQUIPMENT AND INDISPENSABLE INGREDIENTS

Now that you know what you've got going for you (hair and eye color), you're ready to get going—but not so fast! You have to have the right equipment and ingredients on hand, so when the mood strikes you, your personal salon is "open for business." You can't cook up incredible color without the right stuff.

When you've gathered up everything you need, put it all together in a color kit, so you always have what you need neatly organized, on hand, and ready to go! To begin, you will need a metal toolbox (how about one in fire-engine red?), a sturdy canvas bag, or even a stylish storage container to hold your color tools. It is imperative to keep your materials organized and in one designated place so that no one accidentally uses your highlighting toothbrush to scrub up his or her pearly whites. Make sure to keep your kit out of reach of little hands or paws. I keep my at-home color kit in my bathroom cabinet under the sink with a child-safety lock on the cupboard door.

I cannot stress enough how important it is to be organized. There is a precision to color mixing that requires and is, in fact, enhanced by neatness and order. You will find everything you need when you need it in a frustration-free environment! Even extra conditioners from color boxes, leftover tints (clearly marked, of course), and spare gloves (recycling is advised) can be kept in your kit. Small and large Ziploc freezer bags work wonders to keep little things from escaping to the black hole at the bottom of your bag or box. You can use a Sharpie if necessary to mark your bags. I do.

With few exceptions every item on the following lists can be found easily and inexpensively in a drugstore, discount retailer, or supermarket. See the Resource section for places to buy specialty products or websites to order products online (if you simply cannot get to the store).

HERE'S WHAT YOU NEED

Equipment

BATH TOWELS. Two old but clean bath towels are necessary. One should be used to wrap around your neck when you're applying the color mix, to catch and absorb any drips. The other should be used for blotting and drying your hair after shampooing (no matter how well you rinse, it's possible that there's a little residual color left in your hair that could transfer to your towels, so it is best to use a clean one that you don't care about).

BUTTERFLY CLIPS. Depending on how long your hair is, you should have at least six to twelve medium-sized clips. Butterfly clips can often be purchased in sets. They are great for holding hair in place when applying highlights or panels, or even to secure a towel around your neck.

CHILDREN'S MEDICINE DROPPER. This can be found in the baby or infant section of a drug store or pharmacy. Be sure the dropper has measurements in millimeters and ounces, including 1 millimeter up to one teaspoon or five milliliters and up to two ounces. (I prefer two to three so rinsing them is not an issue.)

COTTON BALLS. They are great for cleaning up, blotting drops and spills, and performing some applications.

COTTON SWABS. Q-tip is a good brand. These are useful for applying product and blotting up tiny mistakes.

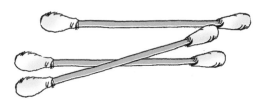

DETANGLE COMB. Detangling combs have wide teeth and a short handle. They are excellent for combing out wet hair after shampooing, combing color through thick hair during the color process, and after towel drying. Hair should always be combed out before applying any color or highlights. Messy hair will result in uneven color. (Do not forget to rinse clean between each color application.)

HAIRBRUSH. A good hairbrush is essential to keeping your locks and scalp in top shape and for brushing out your hair before dampening it for color.

HAIR DRYER. Any good 1200- or 1500-watt hair dryer will do. Unless you blow-dry your hair every day, get whatever is affordable. You don't need anything fancy or expensive. However, if you want to treat yourself to a new dryer, ionic hair dryers are the newest available. These hair dryers break down water for deep moisture penetration, smooth hair surface for silky shine, eliminate static, and deliver

heat gently and quickly. If your hair dryer does not come with a diffuser attachment, don't worry. You can make a great one with a clean white cotton sock (see page 25).

HEAVY-DUTY FOIL. Reynolds Wrap is a good brand. Please make sure it is heavy-duty foil—regular foil will tear! Foil is used for keeping color in or out, depending on the recipe. You *can* reuse foil if you want, but it *must* be rinsed completely and flattened out so that it is iron straight. If it is crumpled or wrinkled in the least, it can create a marbleized effect on your hair, which is something I guarantee you do not want.

ASSORTED LASH, BROW, AND EYE MAKEUP BRUSHES. Use the lash and brow brushes for precise application of brow color. They are great! If you cannot find lash and brow brushes, an eye shadow brush will work in a pinch for brown or darker brows because precision is not as much of a consideration as for lighter brows. An eyeliner brush is the ultimate tool for those of you who have blonde or gray brows. It is the best tool because it is stiffer and allows for a more refined, precise application of color. Mascara brushes also work well for applying color to brows. In fact, the L'Oréal highlighting kits include them, and they are great for applying soft brow color. Rinse and dry brushes thoroughly after each use; they will last a long time.

MEASURING CUPS. Plastic, please; containers can get slippery. Keep sizes from ⅛ cup to 1 cup in your color kit.

MEASURING SPOONS. Standard plastic set with ⅛, ¼, ½ and 1 teaspoon and 1 tablespoon.

MEDIUM-SIZE PLASTIC BOWL. The four-cup size gives you plenty of room to mix and work. Do not use a metal bowl because the metal may react with the color chemicals. I use plastic in the salon because they do not break (believe me, they do fall).

NAIL FILE OR EMERY BOARD. This is a perfect tool for smoothing out dark tints from nails. Even though you will wear gloves when applying color, stains do happen. They are inevitable. Simply smooth out the stain with the emery board or file. If you really have a stubborn problem, take the finest grade of sandpaper you can find and slowly but surely buff the stain off your nails. When I was younger, I did not wear gloves when I colored hair, and people would ask me what I did for a living. I would joke that I was a mechanic because of my almost-permanently stained fingertips! So if you're an at-home hair colorist who isn't into wearing gloves, you can also say you were working on your Harley or Maserati over the weekend!

NOTEBOOK AND PEN. Keep a small notebook handy in your color kit to record recipes, the date you applied them, and the results, so you have an exact record the next time you color or highlight.

PERM KIT. The solution in a perm kit is used by virgin Tab 11s to pre-soften hair before coloring. See Chapter 3 for the pre-softening technique. It couldn't be easier!

PLASTIC FUNNELS. Use them to easily and neatly add liquid to bottles.

2 PLASTIC BOTTLES WITH APPLICATOR-TIP CAPS. You can buy these in drugstores and grocery stores. They are often found in the travel-size product section in travel-container kits. They are perfect for homemade conditioning concoctions or for applying olive oil conditioners to your hair. Try to get one 1-ounce size and one 4- to 8-ounce size.

PLASTIC WRAP. This is useful for wrapping hair to keep moisture and heat in and creating a barrier between your hair and your skin.

RATTAIL COMB. This comb, with its narrow teeth and long, pointy handle, is perfect for creating perfect parts, and, on fine hair, combing color through hair for perfect results.

SPRAY BOTTLE. A small (4- to 6-ounce) spray bottle for water.

THREE-WAY MIRROR. This is your best bet for getting a good all-around look at your hair-color application, especially if you are working alone. A three-way mirror is also useful for styling.

TIMER, SIXTY MINUTES, WITH A LOUD BELL OR BUZZER. A timer is essential when coloring hair. You must leave various applications on for specified times, and the most accurate way to measure time is to use a kitchen timer. It can be manual or electronic, but it must have a loud signal, so you can hear it if you have wandered away from your work area. It is unlikely that you will leave many mixtures on your hair for an hour or more, but a sixty-minute timer is standard. Make sure it works before you use it!

TOOTHBRUSHES. Two, Oral-B brand, soft, with elongated tips. This is the ultimate highlighting tool for adding threads and subtle highlights to your hair.

TONGUE BRUSH. Tung brand tongue brushes are the best tools for adding ribbons of color and giving texture to the ends of your hair. Their 1-inch round head and long handle make them ideal. And Tung brushes are necessities for the thread, thread, and ribbon technique (TTR) used throughout the book and shown in Chapter 3, page 48. You can find them in the toothpaste aisle of your favorite drugstore, usually for less than $5.

WATER JUG. An empty 2.5-gallon jug, which you will cut up to create a paddle to keep hair in place when pulling color or highlights through to your ends.

WHISK. A 4-inch metal whisk works well for blending color mixtures that need to be thickened with cornstarch.

WHITE WASHCLOTHS. Always have 4 or 5 clean cotton terry washcloths on hand for cleaning up, blotting, and wiping hands.

WOMAN'S WHITE COTTON WORKOUT SOCK. This is a perfect use for the sock that has lost its mate to sock heaven. It works remarkably well as a diffuser! Put it on the end of your hair dryer. It works by slowing down the velocity of the heat coming from the dryer and by allowing the warmth to hover in one area, to speed up highlight results.

WORK SHIRT. An old button-down shirt is the perfect "outfit" for coloring hair because it is so easily removed when you are wetting, rinsing, and washing before, between, and after

PADDLE PRO

You can apply color like a pro by making a simple paddle out of an empty plastic water-jug . This simple paddle is the perfect tool for placing underneath strands of hair as you highlight them. It keeps the dye off of underlying hair while providing a firm surface on which to "paint" the strands. First, cut off the handle at the top. Then, at the bottom of the handle, make a cut with a sharp knife (be careful), so you can fit sharp scissors in the hole. Cut around the handle in a shovel shape, about three and a half inches long and four inches wide at the straight end of the shovel.

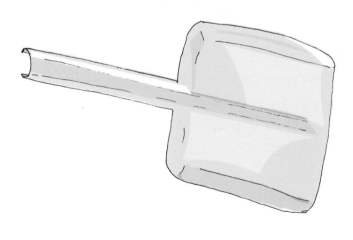

color steps. Think about it: An old T-shirt has to go up and over your head—not convenient when your hair is wet with a color mixture. If you don't have a well-worn one of your own, "borrow" one from the nearest man in your life! When he sees your lustrous locks, he won't mind a bit!

YOYETTES. They are also known as straight or alligator clips because they open like an alligator's mouth and tightly grab on to a section of hair. You can find yoyettes under this name in the hair supply aisle of your favorite drugstore. They are perfect for holding even fine hair in place when applying highlights and lowlights.

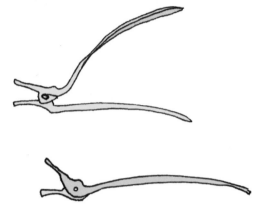

Ingredients

AEROSOL HAIR SPRAY. Spray hair spray liberally onto a cotton ball, and use it to wipe off your hairline immediately after a color application.

CARMEX LIP BALM. This is an essential ingredient for preparing a triple-barrier cream to protect your skin (see Chapter 3). It is the *best* brand out there and available at most drugstores.

CONDITIONER. The conditioners that come in hair-color kits are great, and you should save any extra tubes in your hair-color kit. It is also worth having a good, light conditioner on hand for in-between process shampooing. One of my favorites is Garnier Fructis conditioner. I also love L'Oréal Color Vive conditioners for all types of hair. Pantene conditioners also work well.

CORNSTARCH. This common thickener can be added to highlighting tints to make them easier to work with.

DIFFUSER. A plastic attachment for your dryer that's used to slow down the strong air current and take advantage of its heat benefits.

DISTILLED WATER. This pure water is used as a replacement in recipes that call for a reduced amount of developer in a mixture. I recommend distilled water because all solids and salts have been removed, so there is little chance for unwanted reactions when it is mixed with tints. You can certainly use filtered tap water in a pinch, but a gallon of distilled water is readily available in grocery stores, usually for less than eighty cents.

ESSENTIAL OILS. This is an optional ingredient because you can't always find essential oils in a drugstore. You can find them in soap-making shops and very often in the soap-making aisle of craft stores, such as Michael's. Two drops of your favorite scented essential oil (lavender is lovely in hair) added to two ounces of jojoba oil makes for an effective and relaxing once-a-week conditioning treatment. Just massage the oil into a dry scalp and hair, and leave it in for 2 hours or, even better, overnight and then shampoo it out. See the Resources section (page 289) for additional suppliers of oils and Chapter 12 for more recipes using oils.

40-VOLUME PEROXIDE. This is a product that you must purchase at a beauty-supply store or online. For example, Sally's Beauty Supply is a nationwide chain that sells it. Check the Resource section on page 289 for additional sources. The peroxide in most at-home hair kits is 20-volume. The 40-volume peroxide kicks up many highlights and makes a big difference in the results of the highlighting recipes, especially for the vanilla ribbon recipes. You usually need just a capful from the small bottle, which is why I call it a "cheater cap." In fact, any recipe that calls for peroxide bleach benefits from the addition of a cheater cap.

LEMON OR LEMON JUICE. Keep lemons or a bottle of lemon juice in the refrigerator until you need to use them. Some lemon juice under your nails helps clean them after coloring. Cut a lemon in half and dip the tips of your nails into one of the halves. The juice gets under the nails and helps remove staining. If you don't have a whole lemon, reconstituted lemon juice also works—put some in a little glass bowl and dip your fingertips in.

LOTION. Nivea and Keri brands are excellent. Used in combination with petroleum jelly and Carmex lip balm, it makes an unbeatable barrier cream between your hair and your hairline. See Chapter 3 for the recipe and application technique.

OLIVE OIL. Olive oil makes a great conditioner by itself or as an additive in some recipes. Any olive oil will do; extra virgin works better, however. Keep it in a tightly sealed bottle in your kit—do not expose it to light for extended periods of time (in other words, don't keep your olive oil on the windowsill). Mayonnaise is an acceptable substitute if you don't have olive oil on hand. You can also use oil to remove highlighting mixture from your hair if you have applied too much. Place the oil in a 1-ounce bottle, squeeze it on the excess mixture, and blend the excess mixture with a Q-tip.

PETROLEUM JELLY. Vaseline is a good brand. Use it to create an effective barrier cream around the hairline to deter drips and prevent staining.

SHAMPOO. Shampoo is also included in some hair-color kits, but I prefer Neutrogena or something very gentle that is used for everyday shampooing. Moisturizing shampoos are also good to have on hand, especially if you have lightened your hair. If you have gone red or brunette, L'Oréal and Pantene make excellent color shampoos for washing in between touch-ups. L'Oréal Paris Colorist Collection includes incredible shampoos for color maintenance. I will note when a Colorist Collection shampoo is recommended for certain recipes. You can find Colorist Collection online or in beautysupply shops because it's not a drugstore brand. Prell is a great shampoo to have on hand for washing out the effects of chlorinated swimming-pool water on your colored hair.

WELLA CHOLESTEROL. This is a great product for deep conditioning and can be used in some of my conditioning concoctions (see page 28). It's an especially good conditioner to enhance moisture and shine if your hair is very dry.

WORKSPACE CONSIDERATIONS

A clean, well-ventilated, and open workspace is just as important as organized materials. I am used to the fumes, but first timers may not be. A window fan helps those who are especially sensitive to odors. The bathroom or kitchen is the best place to set up because generally these rooms have tile or vinyl floors that are easy to clean. (Have you ever tried getting hair dye out of carpeting? It's tough!) I also recommend that you clear off the counter space where you are working, if only temporarily, to make room for your hair-color supplies. If you are very worried about porous tiles, or light-colored floors, or wooden floors, line the floor where you are mixing with old newspaper or a paper or canvas drop cloth. I do not recommend lining the area where you will be walking or standing with a plastic drop cloth or plastic bags because plastic is slippery and you may slip and fall. To protect your countertops, you can cover them with a layer of newspaper and then a plastic bag.

If you want to line the walls around your work area with plastic bags, do so. However, you may not need to if you have glossy porcelain tile because it does not stain (though the grout might absorb a stray spot). I get spots on my kitchen and bathroom cabinets when I am working, and even in the salon, we constantly have to repaint the walls. Stains do happen! Use common sense—when waiting for color to develop, don't sit in an upholstered chair. If you are going to sit down and read the paper, it is important to have a towel over the chair. Finally, make sure you have a tall kitchen garbage can "at your disposal," so you can toss your trash—empty boxes and bottles and throw out old cotton balls as soon as you are finished with them. I like my mom's "clean as you go" philosophy.

Work Area Materials

- **NEWSPAPERS.** Use for lining the floor and countertops where you are working.

- **OLD TOWEL.** Excellent for lining the back of chairs when you are resting while the color mixture works.

- **PAPER OR CANVAS DROP CLOTH.** Another option for lining the floor area. Do *not* use plastic drop cloths on the floor!

- **PLASTIC BAGS.** Good for lining countertops, walls, and the back of your chair, under a bath towel.

- **TRASH CAN LINED WITH A STURDY PLASTIC BAG.** Handy for disposing of trash.

- **WASHABLE CHAIR.** Do not sit in your favorite upholstered chair while you wait for the color to work! Stains inevitably happen!

Now that you have your gear together, let's get coloring.

3.

MASTERING THE ART OF FINE COLOR

CONNOISSEUR TECHNIQUES

This is your primer for every technique used in the book. Many of these are tried-and-true professional procedures, and others are methods I've developed in my salon "test kitchen." Each technique is straightforward, but remember you must practice to perfect your personal technique. Go slow, and color your hair when you feel relaxed. Over time you really will become a pro at applying highlights and working all-over color through your hair. The recipes are designed so that even if you don't apply a highlight perfectly the first time, you will still get a good result. This is hair-coloring 101! So take a deep breath and let's get coloring!

BEFORE YOU BEGIN

You will achieve the best results when you prepare not only your work area but also your hair. Before you get coloring, there are three very important steps you must take. First, you must make sure your hair is in good enough condition to color by performing an elasticity or "rubber band" test. It's an easy method and one that can save you plenty of heartache. Second, perform a strand test to ensure that you have chosen a color you love. Finally, after you have prepared your workspace and gathered your materials, you must take the time to protect your face from stains by using my triple-layer barrier cream.

Elasticity Test

All at-home colorists need to perform what I call a rubber band test before embarking on their color journey. It's especially important if your ends are blonde highlighted, straightened, or permed, and you want to go lighter, or if your ends are multicolored and you want to go darker. Gather a one-inch section of your hair and dampen the ends with water— either by spraying them with a water bottle, or, if your hair is long enough, running them under the kitchen faucet and then blotting them with a towel so they are evenly damp. Then clasp the damp hair tightly between your forefinger and your thumb.

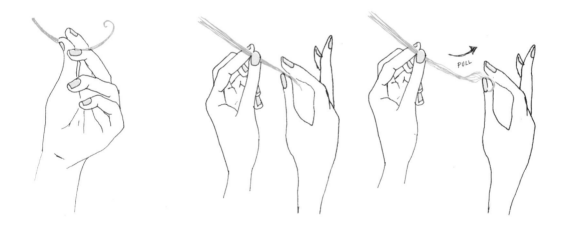

Take your other forefinger and thumb, and pull the hair taut, and then, without letting it go, stop pulling. If the hair bounces back, it is safe to color your hair. If it does not bounce back or if any strands break, you cannot color your hair. It is too damaged and will not take color, or worse, break off. Get your hair cut enough to remove the damaged ends and then try this test again. Otherwise, if you try to color your damaged hair, you may end up getting an unwanted "chemical haircut." In other words, ultradamaged hair will break off if treated with color highlights or bleach. If you are blonde and want to go darker and your hair fails the elasticity test, your ends could turn a hideous green-gray color because only the base part of the color will grab on to the fried ends; the rest of the color will wash away. I don't want to get too scientific, but that's the basic picture, and it's not pretty! It is better to get your hair cut, wait until more of your old color grows out, and get your hair back into good condition. In the meantime, you can use L'Oréal color mousse or Colorist Collection shampoo. Both of these hair products will "color" your hair without adding to the damage.

Strand Tint Test

It is imperative to take a strand test from an obscure area on your head before you color. Some of my clients have actually snipped a "sample" from the very front part of their hairline! Don't do that. Instead, cut a piece of hair about ⅛-inch across and an inch and a half in length from two to three inches behind your ear or two inches above the nape of your neck and under the layer of hair. Or, if your hair is long enough, clip it up and secure the test strand with a yoyette. No one will ever know it's missing or colored.

ANATOMY OF AN AT-HOME COLOR KIT

Differences in packaging aside, all at-home color kits contain the same basic components, which we will use to build the recipes in this book. In every box of single color tint, you will find the following:

- DEVELOPER. This is the creamy, white stuff that activates the tint, so it can color your hair. It comes in a bottle, and tints are added to it. The developer has a built-in "timer" that shortens its shelf life once it is mixed with tint—about thirty to forty minutes. After this amount of time passes, the color power stops.

- TINT OR BLEACH. This is the color itself. Many of the recipes call for more than one tint to be mixed together in the developer. These combinations create shades that meet the requirements of the various tabs. Leftover tint (whatever has not been added to the developers) can be saved for the next use. If you seal it well, it will last several months. When mixing tints together, always use the same brand. It usually does not matter what brand you use (unless one is specified in the recipe) as long as it is consistent within a mixture. For example, use L'Oréal with L'Oréal and Clairol with Clairol. But feel free to experiment with different brands when trying a new recipe. Different types of tints should not be mixed together. That is, use gel tints with gel tints, and liquid tints with liquid tints.

- CONDITIONER. I have found that the conditioners in color boxes are quite good. If you don't use up all of the little tube during a color session (they are fairly generous with this stuff), you can save whatever is left over and use it for touch-up sessions or everyday conditioning.

- GLOVES. These thin rubber gloves can be recycled and used up to three more times, as long as you rinse them thoroughly and coat them inside and out with a fine dusting of baby powder or cornstarch. It's a good idea to save gloves you don't end up using because every once in a while a glove breaks and that spare pair will come in handy.

- DIRECTIONS. Look them over for timing hints. Manufacturers usually recommend leaving the color mix in for thirty minutes, but sometimes they suggest thirty-five minutes.

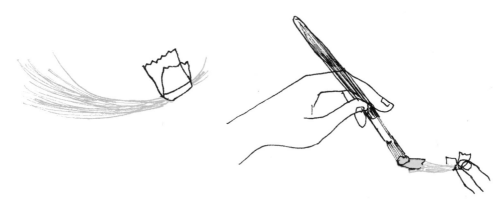

After you've cut (or secured) your sample, use a piece of regular clear tape to hold the hair together, sort of like a tassel. Measure out the recommended amount for the allergy test as given on your color kit and place this in a small plastic bowl or measuring cup. Then measure out your tint or tints in milliliters in the ratio of your chosen recipe and stir them into the developer. For example, a Tab 11 using the Caramel recipe must use 1 teaspoon of developer if using Nice'n Easy, Revlon ColorSilk, or Herbal Essences or 1½ teaspoons for most tube colors, ¾ teaspoon of light golden brown, and ¼ teaspoon of light reddish blonde. (Always refer to the allergy test section on the box for the appropriate developer-to-color ratio for your strand test.) Apply to your test strand, put the sample on or in a piece of plastic wrap, and wait 30 minutes (some manufacturers recommend as long as 45 minutes for resistant gray). Rinse the sample well, shampoo lightly, and blow dry it to see the result (drying should take only minutes for a section this small). Be sure to hold the taped part securely, or the test strand could wash or blow away!

Remember to look at the big picture. Hold the test up to your hairline (where you can see it) and put a white napkin behind it so that you can really see the color. Do you love your recipe test result? If so, prepare your work area and get coloring! If you're not happy with the color, test another sample and use a recipe for another tab in either direction from the one you tested. It is important to follow directions first and then adjust to give personalized results. Your first attempt will be in the ballpark, but it usually takes about three adjustments to score a home run.

MEASURING 101 FOR THE HOME HAIR COLORIST

All the recipes in this book are written with the color container measuring system, which allows you to measure quantities by marking the tubes or liquid containers that hold the color. I do this for my test models and for my clients who use travel kits. Simply take a Sharpie pen and mark the container at intervals to show ½ of the color container, ¼, ⅛, and even ⅟₁₆ (if necessary for the recipe)—see the illustration below.

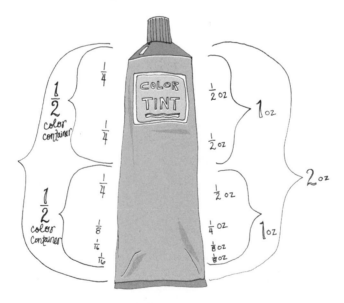

Color containers provide about 2 ounces of tint—some more, some less. What's important to know is that as long as you use the proper proportion, you're set. If a recipe calls for ½ + ¼ of the color container, that's true whether the container is a 1.7-ounce size, 2.4-ounce size, or whatever. Just mark the container and use the right amount of tint.

I also give ounce measurements because some people prefer to measure that way; you can use your regular kitchen measuring cups and spoons. Keep in mind the following equivalents:

4 ounces = ½ cup

3 ounces = ⅜ cup

2 ounces = ¼ cup = 4 tablespoons

1 ounce = ⅛ cup = 2 tablespoons

½ ounce = 1 tablespoon = 3 teaspoons

¼ ounce = ½ tablespoon = 1½ teaspoons

⅛ ounce = ¾ teaspoon (½ + ¼ teaspoon)

1/16 ounce = ⅜ teaspoon (¼ + ⅛ teaspoon)

TRIPLE-BARRIER CREAM

It is *so* important to protect your face from staining while you are coloring your hair. And applying a barrier cream is especially important if you are using any brunette or auburn/red recipes. A barrier is also essential for people with dry skin because their skin has the most potential to stain. In general, oily skin doesn't stain as badly.

The triple barrier is a professional formula that is incredibly easy to apply. Before you begin mixing and applying color, comb your hair away from your face so your hairline is exposed. First, with your fingertips, apply a layer of lotion such as Nivea or Keri around your face near your hairline. Follow that with a thin layer of petroleum jelly. Then, on top of the lotion/petroleum jelly strip, apply a layer of Carmex lip balm. While you're at it, drag some Carmex across your lips—why not give them a treat while you color? I suggest that if you're applying a darker or red colors you should apply triple-barrier cream—especially if covering gray. Saturate a cotton ball with aerosol hair spray and use it to wipe off any excess hair color from around your hairline right after the color application. This step is especially important if you have dry skin or skin that stains easily!

ULTIMATE COLOR TECHNIQUES

These how-to secrets help you achieve incredible results.

Gloss: For Superior Shine

This technique can be used by anyone who wants to add extra shine and richness to existing or natural hair color. Just replace half of the developer with distilled water and add the color to the developer/water mixture. (Be sure to use liquid color.) Leave on for 2 to 20 minutes—less time for lighter colors and more time for darker colors.

Pre-Color: For Superior Coverage

A pre-color is a simple technique that allows light-natural and very gray colors to accept darker tints and, as a result, achieve a richer and more dimensional look. When a pre-color is called for in a recipe, use liquid color. Remove half the developer from the bottle and replace it with distilled water, then add the tint indicated in the recipe. Apply from the scalp through the ends and leave it on for 15 minutes, then proceed as directed. A superlight shampoo with no scalp manipulation, towel drying, comb out, and application of a second color mixture generally follow the pre-color.

Pre-Soften

Pre-softening is recommended for virgin Tab 11 hair (65 percent to 100 percent gray, and hasn't been colored, permed or straightened for four months). Gray hair is tougher to color than other kinds of hair, especially with a liquid color, and to make the tint grab and hold on better, Tab 11s must apply permanent solution from a drugstore-perm kit to their hair. Simply apply Vaseline with strands of cotton around the hairline, saturate the hair with the permanent solution starting about 1¼ inches from the scalp down to the ends, and let it work for 1 minute. Rinse, shampoo, lightly condition for detangling only, comb out, and towel blot your hair. You are ready to continue with your recipe.

A COLOR BY ANY OTHER NAME
IS . . . ABOUT THE SAME COLOR

There are so many brilliant brands on the shelves: L'Oréal, Clairol, Garnier, Nice'n Easy, Revlon, and on and on. Often they number their colors, or they have great names. But they also have a "basic" name that describes the color, and it's right on the box. So when I refer to "light blonde," "light golden brown," or "dark brown," I am referring to a generic color that you can find in pretty much any brand. I do this because I don't know how many different brands your local drugstore carries and I don't want you to be stuck. In fact, often on the side of the box, manufacturers will even compare colors. Revlon is wonderful for this. For example, you will sometimes see something like this on a box: "If you have used Clairol's Brown, you can use this brown . . ." There are times when I recommend a specific brand and color just because I have gotten a great result with it and I want to share my findings with you. That will be noted in the recipe; otherwise, you can use any brand of color indicated in the recipe. I've found that if you want better gray coverage or more deposit for light hair going darker or redder, Garnier Nutrisse, L'Oréal Couleur Experte, L'Oréal Preference, Clairol Hydrience, Revlon Colorist, Revlon ColorSilk, and Nice'n Easy are your best bets. For darker tones going lighter, I think you get better lift with L'Oréal Féria, Revlon High Dimension, and the Garnier Nutrisse blondes. However, as noted earlier, once you pick a brand, be sure to use it consistently throughout a recipe.

Removing Darker than Natural (Tinted) Hair Color

Whether your hair is currently dark or you're not thrilled with the depth of color, simply purchase a box of Color Oops hair color remover and follow the manufacturer's directions. In twenty minutes you'll be back to where you need to be in order to try again, or proceed to your local salon. Please see the Resources section (page 289). I recommend that all home hair colorists keep a box of Color Oops in their tool kit just in case.

Virgin Application—All-Over Color

When you apply an all-over color (as opposed to a highlight or lowlight) to your entire mane, it's called a "virgin application," and the all-over recipe instructions in this book assume this is what you will be doing. All-over color creates a basic canvas to be left as is or as a base for highlights and lowlights, *and* it covers gray. This is the basic technique for all-over application, and it should be used every time you give yourself an all-over color.

1. Prepare your work area by covering the floor and counter; take your kit out and gather your color boxes (as indicated in your recipe).

2. Comb your *dry* hair through to detangle and brush it away from your face.

3. Apply the triple-barrier cream.

4. Mix your color.

5. Follow all-over application instructions in your recipe. Unless otherwise indicated, this usually begins by applying the color mixture 1½ inches down from your scalp to the ends and leaving it on for a designated amount of time (usually between 5 to 10 minutes, as indicated in the recipe) and then applying the remainder of the color from the scalp down 1½ inches to meet the already applied color. All of the mixture you have applied is left on for the rest of the working time, usually another 20 to 25 minutes, for a total working time of 30 to 35 minutes. Comb the color through with your detangle comb to ensure it is evenly distributed.

Retouch

Retouching is done when you need to refresh an all-over recipe. Nothing lasts forever, unfortunately. Color fades, and hair grows out. Retouching entails applying color on the regrowth area, from the scalp to where existing color is, then applying the recipe through the ends for 5 or 10 minutes (if you don't have highlights), or you can try the foil-out technique for texture.

Once you have applied the recipe for your target color, you will begin to do touch-ups in two weeks (for hairline and part) to eight weeks (for a full hair touch-up), depending on your natural color and how quickly your hair grows. If your target color is close to your natural shade, you will be able to wait a little longer, up to eight weeks, before you do a touch-up. If you have gone darker or lighter, you may have to do a touch-up more frequently.

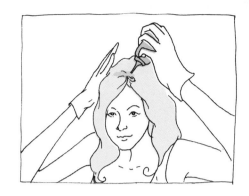

Foil Out

Use the same recipe as you did in your virgin application to maintain your look. If you make an alteration to your recipe (such as kicking up the color by adding more of a lightening or darkening tint, as suggested in some recipes), you follow instructions for an all-over application but reverse the process in that you would apply the recipe from your scalp to 1½ inches down and leave on for 15 to 20 minutes (whatever the recipe indicates), foil out the hair, and then apply the remainder of the recipe through the ends, for the rest of the work time, usually up to 15 minutes.

This is one of my favorite ways to make hair look even better. Go beyond application to the regrowth area and do a foil-out technique using the remainder of your recipe. Here's how.

1. Take ¼- to ½-inch-thick ribbons of hair on your part and around your face and wrap them up individually with a piece of heavy-duty foil. It is good to put conditioner on each section as you wrap it; I do this in the salon. Generally I put two ribbons on one side of the part and three ribbons on the other side, staggered so that they don't line up exactly.

2. Apply the remaining mixture to the hair sections that aren't in foil, starting at the nape of the neck and moving up to the hairline. Comb through evenly, section by section.

3. Set the timer for 15 minutes. If you want deeper or brighter color, start the foil out after the recipe has been on the hair for 10 minutes.

4. When the time is up, rinse your hair with the foil still on it. When the water runs clear, remove the foil, shampoo and condition your hair, towel blot, and style as usual. *Voilà!* You've got subtle, natural highlights and texture without bleaching.

Midprocess Shampoo and Towel Blot

Many of the color-mix recipes contain more than one step and between color steps you must shampoo out color before applying more color. To protect your face and eyes, take some simple precautions when washing color out of your hair. First, your hair should be away from your face. Step under the shower head, making sure you are facing the spray. Tilt your head back and let the color wash out. Refrain from any scalp manipulation. In other words, do not scrub or massage your scalp! Gently work your hands through the hair to move the color out. When the water runs clear, shampoo with a gentle or moisturizing shampoo, rinse well, and give hair a light condition. That means leaving a light conditioner (the one that comes in the color box is fine) on your ends for 4 seconds and then rinsing it off. Conditioning helps with detangling. Towel blot your hair and comb it out before proceeding. However, if you are adding highlights next, hair must always be *completely* dry before you begin the application.

Color Shampoo

When using color-enhancing shampoo, such as Colorist Collection, apply it first to the hair at the nape of your neck, then massage it in moving toward the front of your head and through the ends. This prevents you from having a big splotch of color at the top of your head. It's a good idea to wear gloves when using deep-color shampoos to prevent even the most minimal stains on hands and nails.

Highlighting and Lowlighting

Highlights on the exterior of the hair add warmth and light. Lowlights add richness and dimension from under the top layer of hair. The result looks more natural. That's because in nature, the sun bleaches the outermost portion of your hair.

However, if you color every single strand of your topmost layer of hair a different shade of blonde, the effect is one color—the sun isn't that precise! Avoid overlightening bangs as well. That much one color of blonde around your face ends up washing you out, and it won't look real. Instead, start the highlights in the area just above your fringe or bangs or as a face frame. Always apply highlights and lowlights bottom heavy. In other words, apply more highlight mixture to your ends and don't begin the highlights at your scalp. Instead, start them about 1½ inches down from your scalp.

There are many methods of applying highlights and lowlights, as described below. Use the highlight recipe indicated under your tab recipe. Of course, any highlight recipe in the book can be used on its own, without the accompanying color recipe, on existing hair color. The following is a standard highlight recipe and application technique that you will be referred to in some of the recipes.

STANDARD HIGHLIGHT RECIPE

Newspaper or paper or canvas drop cloth for floor-work area

2 worn bath towels

2 butterfly clips

Button-down work shirt

1 hard stick of butter, or 1-inch piece of citrus rind (from a yellow grapefruit, orange, or lemon), or 1 piece of light caramel, depending on the color you want to match

Clairol Frost'n Tip Dramatic brand highlight kit

1 one-ounce plastic bottle with applicator tip filled with olive oil (see Resources on page 289 to find the bottle)

1 large drop cloth

1 teaspoon (approximately) Vaseline petroleum jelly

10 to 20 cotton swabs (Q-tip is a good one)

2 Oral-B soft toothbrushes with elongated tips

1 Tung brush

1 paddle made from a water jug (see page 25 for instructions)

8 to 12 cotton balls, unrolled, or 8 to 12 cotton strips, slightly pulled apart to form sheets

1 or 2 teaspoons of olive or jojoba oil (this addition helps with the ease of application since at-home kits are so dry)

2 to 4 white washcloths

Plastic wrap

1 plastic bowl (if a container is provided in the kit, use that)

1 hair dryer with diffuser attached, or your cotton-sock diffuser

DIRECTIONS

1. Prepare your work area, place the drop cloth under your feet, and put on your button-down work shirt.

2. Open the Clairol Frost'n Tip box and remove the gloves, developer, powder bleach, and conditioner.

3. Drape a worn bath towel around your neck and secure with a butterfly clip.

4. *If you have bangs or short front layers,* unroll or elongate 3 cotton balls and lay them on top of one another. Pull them again so that they become one thick strand. (You can also purchase ready-made cotton strips online or from a beauty supply store.) Spread a thin layer of petroleum jelly just under your bangs in a strip about 3 inches long and place the cotton strand on top of the jelly, where it will stick like glue. The band will lift your bang section up so that the highlight does not get into your eyes. If you need to build up the cotton hill even more because your bangs are very long, do so until you are satisfied.

If you do not have bangs, spread petroleum jelly lightly on the cotton ball strips and stick to your forehead under your hairline and to the sides of your face in front of your hairline.

5. Mix the bleach, developer, and oil together in the bowl.

6. Put on the gloves that come in the highlight kit, look in the mirror, and growl, "I can do it!"

7. Starting in front of your ear, separate a ¼-inch ribbon, hold it very taut, and lay the ends of your hair on the plastic paddle. With the Oral-B toothbrush, apply the mixture starting from your scalp to the midsection of the ribbon. Then use the Tung brush to apply the bleach mixture on the ribbon starting from your scalp all the way to the ends. If you are retouching new growth on a highlight, follow your old pattern and stop at the old highlight.

8. Wipe your gloves off on washcloths after each stroke is finished. Try to apply the mixture lighter at your scalp and thicker through your ends.

9. Use the plastic bottle with the applicator tip filled with olive oil to "blend" away any unwanted product.

10. Lay additional cotton strips and sheets of plastic wrap over the highlights and start keeping time.

11. Using your hair dryer with a diffuser or sock attached, hover around each section for counts of 20, starting at the ends. Heat helps kick up the highlights. I use this technique in the salon and on house calls, and the results are gorgeous!

12. Leave the mixture on for 5 minutes for subtle sunshine or as a pre-step to all-over color, or up to 50 minutes for more resistant hair or very bright sunshine. Look to see if the highlight matches the inside of a citrus rind for palest highlights, a stick of butter for a lemon blossom color, or a caramel candy for caramel highlights.

13. Rinse and shampoo your hair. *Voilà!* Excuse me, did you just get back from the islands?

HIGHLIGHT TIMING TIP

The longer bleach stays on the hair, the lighter the hair gets, so remember—for subtle lights, start checking at 5 minutes, and for lighter lights, start at 10 minutes.

Threading

This hair-painting technique duplicates delicate embroidery thread, resulting in a very subtle effect. Use the basic highlighting recipe just detailed or another highlighting recipe for

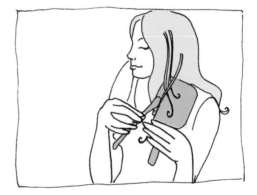

your tab to paint fine threads at your part and around your face. Take your Oral-B toothbrush, dip its tips in the highlight mixture, and paint or brush the highlight mixture on about 5 to 7 strands of hair, starting at your scalp and brushing all the way down to the ends of your hair. Use your homemade water-jug paddle to isolate ends of your hair in the back of your head, and apply the highlight mixture using the Tung brush. Place threads randomly around your head, staggered on either side of your favorite part. If this is a first-time endeavor or if you are going for highlights that look like you've spent some time at the beach, use the Tung brush to pull some of the mixture through your ends.

Ribboning

Use the basic highlighting recipe previously detailed or a highlighting recipe for your tab to paint ribbons flat on your head, without isolating the hair with a paddle or yoyette. Simply take your Oral-B toothbrush, dip it into the highlight mixture, and paint or brush the highlight mixture down your hair starting at your scalp. The top of the ribbon should be ¼- to 1-inch wide and gradually widen to ¾-inch at the ends of the hair. Use your homemade water-jug paddle to measure (it's 4 inches wide—you can mark out ¼-, ½-, and 1-inch intervals on it with an indelible marker for easy reference). Each ribbon should be 1 or 2 inches from one another and staggered on either side of your part. If your hair is faded, this is an easy and highly effective way to refresh it.

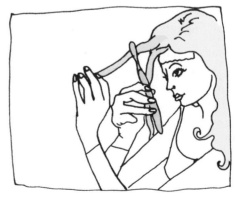

Thread, Thread, and Ribbon Technique (TTR Technique)

This is an amazing way to create natural-looking subtle highlights all over your hair. It will change the way you do highlights forever! An Oral-B toothbrush is needed, as is a Tung brand tongue brush.

MATERIALS

Oral-B toothbrush, soft, with an elongated end

Tung brush

Homemade plastic water-jug paddle (see page 25 for instructions on how to create this)

Plastic wrap

to 12 cotton strips or unrolled cotton balls

DIRECTIONS

1. Use any highlight mixture. Pick up ½-inch ribbon of hair closest to the right side of your head. Hold it at its midsection, laying the ends on your paddle. Dip the Oral-B toothbrush into the highlight mixture and slide it in on either side of the ribbon as if you were drawing a border.

2. Put down the Oral-B toothbrush and pick up your Tung brush. Dip it into the same highlight mixture, and start pulling the product through the ribbon ¼-inch down from where the borders you made end. Continue pulling it down the center of the ribbon without adding any new product. Lift the Tung brush up every ¼ of an inch and overlap the product at the end slightly, feathering it down until you reach the end of the ribbon. The product will be lighter at the start and very heavy at the ends. Lay cotton strands over each TTR section (this keeps the plastic wrap from smooshing the product and giving you spots). Continue highlighting the ribbons with about ½ inch between each ribbon all the way around your head until you reach the other side.

3. When you're done, make a custom plastic heat cap with two sections of 14-inch-long plastic wrap. Lay one piece over your hair front to back, and then lay the other piece side to side over your hair. Plastic sticks to itself very well, so softly lay the "cap" around your head, and it will stay put. The plastic retains the natural heat from your head and keeps the product from drying out too much. It also forms a barrier between your hair and hair dryer with a diffuser or diffuser sock.

Balayage

In French, the *balayage* technique means to sweep with light. Sounds lovely, doesn't it? It is! It gives you the freedom to put color where you want it. How? Dip your Oral-B toothbrush, already explained in the equipment list, into your highlight mixture and use it to brush freeform highlights where you want them. Use it to create face-framing highlights, which add sparkle and warmth around your face. Or use the same

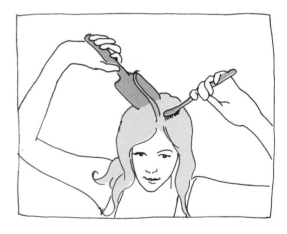

technique for a face-framing lowlight by painting pieces dark to create depth and richness around your face, and pull them through the ends using a Tung brush. Use your homemade water-jug paddle to hold the ends.

Refine the Light

Highlighting does not have to be perfectly lined up (it won't look natural). Try to apply them as smoothly as possible, but if highlights get out of hand and you get spots of highlight mixture where you don't want them—or a highlight ribbon gets too wide—you can remove excess highlight mixture with a Q-tip dipped in oil (olive, jojoba, or almond will do), or you can fill a little 1-ounce bottle with oil and delete the highlight. It can help you refine your strokes. You can also use the oil-dipped Q-tips to remove mixture that has gotten on your skin, especially around the hairline. I do this for all my clients in the salon, and it's fantastic!

The Smudge

This is one of my key techniques and is great for women, men, and teenagers. Simply mix your all-over-color recipe and apply as directed. Leave it on for just 3 to 4 minutes, and then remove it with a rinse and a gentle shampoo. Like Pledge on woodwork, it gives your hair a little something extra by kicking up the color a tiny bit, and depending on the recipe, the result will be either redder or more neutralized. It is not used for gray coverage, but it does help to blend the gray away.

This is also an amazing technique for a "salon color." You can stay loyal to your colorist while using this method around the hairline and part in between salon visits. It takes away any shadows created by regrowth and makes color feel fresh and brand new until your next appointment.

THE AT-HOME COLOR CONSULTATION

STEP ONE: DETERMINE YOUR EYE COLOR

Stand in front of the mirror and take a close look at your eyes. Which color or tone most closely matches your eye color? By paying attention and accentuating the warm or cool tones in your eyes, you'll achieve flattering and natural-looking highlights. Yet don't be afraid to play. Break the rules with a certain color if it makes you feel your best, and if you do, make sure to modify your skin tone with makeup colors that pull your look together. Light skin tone with Warm eye color with licorice—why not? Dark skin tone with Cool eye color with caramel might be dreamy—who says you can't be who you want to be? Remember, the basics are just the beginning; after that the world of color is yours to customize!

Check out page 9 for further help in determining your true eye color—it's more than just brown, blue, or green!

DARK WITH WARM: Girls with glints of gold in their eyes look pretty in warm brown lowlights and gold and coppery highlights. Caramel and Buttered Lights would be very delicious.

LIGHT WITH WARM: Beachy babes with sunny flecks of yellow or tiger's-eye brown in their eyes are breath-taking in pale highlight versions of Buttered Lights and Caramel threads running throughout their manes.

DARK WITH COOL: Beauties with flashes of deep blue, dark brown, ivy green, or gray flecks will find Cream Swirl highlights enticing, or how about some hints of red or Raspberry Lights for high-drama lowlights?

LIGHT WITH COOL: When your eyes twinkle with pale blue tones like sky, lavender, sea foam, and blue-green, you'll look luscious in Dreamsicle, Cream Swirl, and Raspberry Lights.

STEP TWO: DETERMINE YOUR NATURAL HAIR COLOR

Find the tab that most closely resembles your natural shade. Look to the hair near the scalp for the most accurate assessment. Be sure to examine your hair in strong natural light—preferably between 10 AM and 2 PM—or under a GE Reveal lightbulb for extra clarity. Sometimes it is a process of elimination; when in doubt, always choose the lighter tab. You can always intensify your recipe for better coverage next time. Tabs 2 and 3 can look almost the same even to professionally trained colorists, but they usually follow the same recipe anyway. Ashes are murkier in tone and have a slight greenish-blue cast to them. The neutral (non-ash) colors are a little warmer (redder) in tone.

Read page 7 for further instructions on how to pick your natural hair color tab.

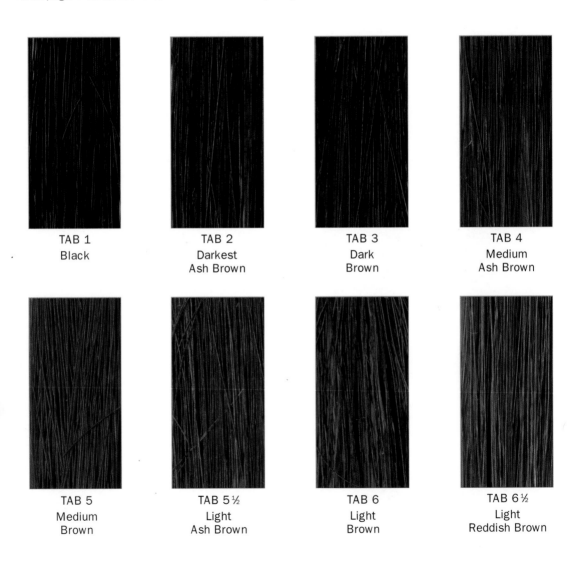

TAB 1
Black

TAB 2
Darkest
Ash Brown

TAB 3
Dark
Brown

TAB 4
Medium
Ash Brown

TAB 5
Medium
Brown

TAB 5½
Light
Ash Brown

TAB 6
Light
Brown

TAB 6½
Light
Reddish Brown

TAB 7
Dark
Ash Blonde

TAB 7 ½
Dark
Blonde

TAB 8
Medium
Blonde

TAB 9
up to 35% white

TAB 10
35%–65% white

TAB 11
65%–100% white

STEP THREE: FIND YOUR NEW DREAM COLOR!

Are you going more golden? Is red your secret beauty fantasy? Have fun exploring the possibilities.

Check out the following pages to find the photo that most closely resembles your desired shade. Remember that not all colors will work for you. Be sure to read Chapter 1 for further instructions on how to pick your perfect color. You want to enhance what Mother Nature gave you; you don't want to pick a color that won't work with your eyes and natural hair color.

The photos that follow are meant simply to give you an idea of what colors you can achieve with the recipes in this book. Please note that I did not necessarily color the hair of the celebrities shown, nor am I suggesting that they all color their hair or that they endorse my work. In fact, the lovely Dita Von Teese (see next page) always does her own hair coloring, hair styling, and makeup—an inspiration to us all!

Dita Von Teese is ethereal in Licorice (FOR ALL TABS)

Lucy Liu is steamy in Espresso (FOR ALL TABS)

Espresso with Raspberry Lights add to Halle Berry's flawless natural beauty (FOR TABS 1–6, 7–11)

Salma Hayek is delicious in Dark Chocolate (FOR TABS 1–6)

Sandra Bullock is sultry in Dark Chocolate with Deep Caramel Ribbons (FOR ALL TABS)

Selma Blair looks dreamy in Milk Chocolate (FOR ALL TABS)

Penélope Cruz is stunning with delectable Cream Swirled Milk Chocolate locks (FOR ALL TABS)

Eva Mendez is quite the vixen in Cherry Cola (FOR TABS 2–11)

Julia Roberts spices things up with Toasted Cinnamon (FOR ALL TABS)

Amanda Peet looks enchanting in Toasted Cinnamon with Buttered Lights (FOR ALL TABS)

Keira Knightly, scrumptious in Deep Caramel (FOR ALL TABS)

Fergie is foxy in Deep Caramel with Vanilla Ribbons (FOR ALL TABS)

Jennifer Aniston looks sweet and sexy in Toasted Wheat with Vanilla Threads (FOR TABS 5 ½–11)

Tia Carrere looks rich and gorgeous in Smoky Plum Wine (FOR ALL TABS)

Drew Barrymore, exquisite and captivating in Merlot (FOR TABS 3–11)

Marcia Cross, vibrant in Spiced Persimmon (FOR ALL TABS)

Julianne Moore is fiery in Gingered Toffee (FOR TABS 4–11)

Heather Graham glows in Apricot (FOR TABS 1–5 IN SHORT HAIR ONLY, 5½–11 IN LONGER HAIR

Mary-Kate Olsen is fresh and beautiful in Amber Honey Dream with Lemon Blossom Ribbons (FOR TABS 5 ½–11)

Cameron Diaz is sun-kissed in Vanilla Dreamsicle (FOR TABS 5 ½, 6, 7–11)

Yummy—that's Jessica Simpson in Vanilla'd Butterscotch (FOR TABS 5½–8)

Breathtaking Marilyn Monroe in Brandied Fig (FOR TABS 5 ½–8)

Pink is workin' it in Peekaboo Punk (FOR ALL TABS)

Avril Lavigne is divine with Dip Tipping (FOR ALL TABS)

Mischa Barton is babe-a-licious in The Wrap (FOR ALL TABS)

Brad Pitt is so handsome in Frosted Secret Sauce (FOR ALL TABS)

Always debonair, George Clooney with Delete the Gray (FOR ALL TABS)

Johnny Depp is pure rebel in his Surf Punk (FOR ALL TABS)

Tommy Lee—the ultimate Rock in Rock Star (FOR ALL TABS)

PART II

The Recipes

4.

SULTRY, DEEP BRUNETTES

Ebonize your brown—it's beautiful! If you've always dreamed of being dark and mysterious, these recipes will help you get there. They create the deepest, most delicious hues of brunette. The target colors are

- **LICORICE**

- **ESPRESSO**

- **ESPRESSO WITH RASPBERRY LIGHTS**

- **DARK CHOCOLATE**

- **DARK CHOCOLATE WITH DEEP CARAMEL RIBBONS**

NOTE: All recipes give instructions for virgin hair. For instructions for retouching or any other technique, refer to Chapter 3.

NOTE: If you have very thick, long hair (shoulder length or longer), double all recipes.

READ ME! PREPARATION IS EVERYTHING

Before you begin any recipe, read through the list of ingredients, get out your color kit, and prepare your work area. Clear off the countertop you will be working at and spread out your materials. Detangle your hair without manipulating your scalp. Put on your work shirt and wrap a towel around your neck and secure with a butterfly clip to catch any potential stains. Apply the triple-barrier cream around your hairline (page 37). This step is especially important for brunette recipes. Cover the floor with a drop cloth or old newspapers. (Remember, no plastic under foot!) Be sure to have your timer out and ready to be set. *Wear kit-provided gloves when applying color!* You can recycle gloves by rinsing them inside and out, patting them dry, and sprinkling some baby powder or cornstarch inside for easy slip-on next time. If your hair is thick or below shoulder length, have two or three boxes of your main color on hand. You don't want to run out!

NOTE: As long as you don't introduce the developer to the tint, you can pre-mix colors and seal them tightly. Use them within six months to a year. However, once the tint is mixed with the developer, you have about thirty minutes of work time. Anything left over must be discarded.

NOTE: If you have very short hair, use the Frosted Secret Sauce highlight technique found in Chapter 10, on men's hair color, page 247.

LICORICE

Luscious Licorice is as glossy and smooth as the name implies. Hair color has never been sweeter! My favorite brands for Licorice looks are Nice'n Easy 124, Revlon Colorist, L'Oréal Color Pulse Mousse Electric Black 10, and Garnier 100% Color #210.

Gloss for Tabs 1 through 3

Tab 1 ladies are already at Licorice, but they can use this simple recipe to kick in shine and fulfill their craving for this sweet treat. Tabs 2 and 3, who are looking for a less permanent change, can use it to get that dramatic Licorice hue.

MATERIALS

Triple-barrier cream (page 37)

1 box blue black tint

Measuring cups and spoons

Distilled water

8 cotton balls

Aerosol hair spray

Timer

Moisturizing shampoo

1 box L'Oréal Paris Color Pulse Concentrated Color Mousse, Electric Black 10

DIRECTIONS

1. Detangle your hair and comb it out.

2. Apply the triple-barrier cream and put on gloves.

3. Snip off the applicator tip of the developer and take off the cap.

4. Using a measuring cup, remove 1 ounce (½ color container) from the kit's developer bottle. Discard it. (It can be safely rinsed down the drain.)

5. Using a measuring cup, add 1 ounce (½ color container) distilled water to the developer bottle.

6. Add the entire bottle of tint to the developer and mix well until the two are combined, holding a gloved finger over the open tip.

7. Apply the mixture to your hair, starting from your scalp to the ends and comb through your hair so the mixture is evenly and completely distributed.

8. As soon as you are done with the application, saturate cotton balls with aerosol hair spray as needed and wipe away the triple-barrier cream in the direction of the hairline to avoid creating a harsh line. This ensures you will not stain your skin.

9. Set the timer for 30 minutes and relax!

10. When the timer stops, rinse and shampoo your hair.

11. Squeeze excess water from your hair.

12. Use a moisturizing shampoo for everyday cleansing, and follow with the conditioner of your choice.

13. Use L'Oréal Electric Black every two weeks for the ultimate Licorice shine! If your hair is permed or straightened, use it with no gloss.

TRIPLE-BARRIER CREAM—A REMINDER

First, with your fingertips, apply a layer of lotion such as Nivea or Keri brands around your face at your hairline. Follow that with a layer of petroleum jelly. Apply a layer of Carmex lip balm over the lotion and petroleum jelly. Never forget to apply the barrier cream when applying very dark colors!

Tabs 2 through 6½

Tabs 2 and 3 should use this recipe for a more permanent Licorice look. Tabs 4 through 6½ will need to use this recipe in order to create the depth needed for true Licorice.

MATERIALS

Triple-barrier cream

1 box blue black tint

cotton balls

Aerosol hair spray

Shampoo and conditioner

1 box L'Oréal Paris Color Pulse Concentrated Color Mousse, Electric Black 10

DIRECTIONS

1. Apply the triple-barrier cream and put on gloves.

2. Snip off the applicator tip of the developer and take off the cap.

3. Mix the black color as per the manufacturer's directions.

4. Add the entire bottle of tint to the developer and rock the bottle gently back and forth until the two are combined, holding a gloved finger over the open tip.

5. Apply the mixture to your hair, starting from your scalp to the ends and comb through it so the mixture is evenly and completely distributed.

6. As soon as you are done with the application, saturate cotton balls with aerosol hair spray as needed and wipe away the triple-barrier cream in the direction of the hairline to avoid creating a harsh line. This ensures you will not stain your skin.

7. Set the timer for 30 minutes and relax!

8. When the timer stops, rinse and shampoo your hair gently with no scalp manipulation.

9. Towel blot your hair and apply L'Oréal Electric Black Mousse to your hair. Comb it into your hair and leave it in to work for 10 minutes.

10. Rinse until the water runs clear and shampoo your scalp well. Condition with the kit's conditioner or a light detangler.

Tabs 7 and 8

MATERIALS

1 box light auburn (Nice'n Easy, Revlon ColorSilk, Clairol Herbal Essences)

Triple-barrier cream

1 box L'Oréal Paris Color Pulse Concentrated Color Mousse, Electric Black 10

1 box blue black

Ardell Gray Magic

8 cotton balls

Aerosol hair spray

Timer

Cleansing shampoo

DIRECTIONS

1. Pre-color with the light auburn tint (see the technique on page 38), including towel blotting and combing out your damp hair.

2. Apply the triple-barrier cream and put on gloves.

3. Snip off the applicator tip of the developer and remove the cap.

4. Mix the black color as per the manufacturer's directions and add 3 drops Gray Magic.

5. Add the entire bottle of tint to the developer and rock the bottle gently back and forth until the two are combined, holding a gloved finger over the open tip.

6. Apply the mixture to your hair, starting from your scalp to the ends and comb it through so the mixture is evenly and completely distributed.

7. As soon as you are done with the application, saturate cotton balls with aerosol hair spray as needed and wipe away the triple-barrier cream in the direction of the hairline to avoid creating a harsh line. This ensures you will not stain your skin.

8. Set your timer for 35 minutes and relax!

9. When the timer stops, rinse and shampoo your hair. Towel blot and detangle.

10. Put on gloves and apply L'Oréal Electric Black Mousse for 10 minutes. Rinse until the water runs clear, shampoo scalp gently and condition through the ends, and style as usual.

11. Use L'Oréal Electric Black Mousse every two weeks.

Tab 9

Use the recipe for your pre-gray tab.

Tab 10

If you're less white, use the recipe for your pre-gray tab. If you're more white, use the Tab 11 recipe.

Tab 11

MATERIALS

Ardell Gray Magic

1 perm kit

Triple-barrier cream

1 box L'Oréal Paris Color Pulse Concentrated Color Mousse, Electric Black 10

1 box copper

1 box blue black

Distilled water

8 cotton balls

Aerosol hair spray

Timer

Cleansing shampoo

1. If this is a virgin application, pre-soften your hair. If you have sensitive skin (scalp), do this one day before you color. Apply the permanent solution from the drugstore perm kit to your hair for one minute, then rinse it out and towel dry. Do this by simply applying the solution to your hair starting about 1¼ inches from your scalp down to the ends. Rinse, shampoo, and comb out your hair. You are ready to continue with your recipe.

2. If this is a virgin application, you must pre-color your hair with a deep copper tint. Mix as instructed and apply it starting ½ inch from your scalp down through the ends and leave it in for 15 minutes. Shampoo your hair with no scalp manipulation. Towel blot your hair and detangle.

3. Apply the triple-barrier cream.

4. Snip off the applicator tip of the developer and remove the cap.

5. Add the tint and 3 drops of Gray Magic to the developer and shake the bottle gently until the materials are combined, holding a gloved finger over the open tip.

6. Apply the mixture to your hair, starting from your scalp to the ends, and comb through your hair so the mixture is evenly and completely distributed.

7. As soon as you are done with the application, saturate cotton balls with aerosol hair spray as needed and wipe away the triple-barrier cream in the direction of the hairline to avoid creating a harsh line. This ensures you will not stain your skin.

8. Set the timer for 35 minutes and relax!

9. When the timer stops, rinse out, shampoo your hair gently, and towel blot.

10. Apply the L'Oréal Electric Black Mousse, comb it through your hair, and let it work for 10 minutes.

11. Shampoo scalp gently and rinse until the water runs clear.

12. Use a color-friendly shampoo for everyday cleansing to extend your color. Use L'Oréal Electric Black Mousse every two weeks for maximum depth of color.

ESPRESSO

This is a soft version of Licorice. Just like a steaming cup of espresso, this hair color has great depth and shine.

Tab 1

MATERIALS

Triple-barrier cream

1 box dark brown

8 cotton balls

Aerosol hair spray

Timer

Cleansing shampoo

L'Oréal Paris Colorist Collection Cocoa Bean Shampoo*

*Available at beauty supply stores and online (see the Resources section on page 289)

DIRECTIONS

1. Detangle your hair and comb it away from your face.

2. Apply the triple-barrier cream and put on gloves.

3. Snip off the applicator tip of the developer and remove the cap.

4. Add ¾ bottle of tint to the developer and shake the bottle gently until the two are combined, holding a gloved finger over the open tip.

5. Apply the mixture starting from 1½ inches away from your scalp down to the ends and set timer for 12 minutes.

6. Apply the remainder of the tint mixture starting from your scalp to 1½ inches down to meet the area where the tint has already been applied. Comb the mixture through your hair from your scalp to the ends to make sure it is distributed evenly.

7. As soon as you are done with the application, saturate cotton balls with aerosol hair spray as needed and wipe away the triple-barrier cream in the direction of the hairline to avoid creating a harsh line. This ensures you will not stain your skin.

8. Set timer for an additional 3 minutes (approximately 15 minutes total work time).

9. When time is up, rinse your hair and shampoo it with no scalp manipulation.

10. Towel blot your hair and apply the conditioner provided by the manufacturer. Work it in down to the ends of your hair and leave it to work for 10 minutes.

11. Rinse and style as usual. Use Colorist Collection Cocoa Bean Shampoo every third shampoo to extend your color.

Tabs 2 and 3

MATERIALS

Triple-barrier cream

Measuring cups and spoons

1 box dark brown

1 box soft black

8 cotton balls

Aerosol hair spray

Timer

Moisturizing shampoo

L'Oréal Paris Colorist Collection Cocoa Bean Shampoo*

*Available at beauty supply stores and online (see the Resources section on page 289)

DIRECTIONS

1. Detangle your hair and comb it away from your face.

2. Apply the triple-barrier cream and put on gloves.

3. Snip off the applicator tip of the developer and remove the cap.

4. Using measuring cups or spoons, add 1¾ ounces plus ⅛ ounce (three ¼ + ⅛ + ¹⁄₁₆ color container) of dark brown tint to the developer.

5. Add ⅛ ounce (¹⁄₁₆ color container) of soft black tint to the developer.

6. Replace the cap of the developer and shake the mixture until it is combined.

7. Set the timer for 30 minutes.

8. Apply the mixture starting from 1½ inches down from your scalp to the ends. Comb through your hair so the mixture is distributed evenly. Leave it in for 5 minutes.

9. When time is up, apply the rest of the mixture starting from your scalp to 1½ inches down.

10. Comb through your hair again to make sure the mixture is distributed evenly.

11. As soon as you are done with the application, saturate cotton balls with aerosol hair spray as needed and wipe away the triple-barrier cream in the direction of the hairline to avoid creating a harsh line. This ensures you will not stain your skin.

12. Let the color work for 20 minutes, until the timer goes off.

13. Rinse your hair and shampoo it without manipulating the scalp. Condition and towel dry it. Style as usual. Use Colorist Collection Cocoa Bean Shampoo every third shampoo to extend your color.

Tab 4

MATERIALS

Triple-barrier cream

Measuring cups and spoons

1 box darkest brown

1 box black

8 cotton balls

Aerosol hair spray

Timer

Moisturizing shampoo

L'Oréal Paris Colorist Collection Cocoa Bean Shampoo*

*Available at beauty supply stores and online (see the Resources section on page 289)

DIRECTIONS

1. Detangle your hair and comb it away from your face.

2. Apply the triple-barrier cream and put on gloves.

3. Snip off the applicator tip of the developer and remove the cap.

4. Using measuring cups or spoons, add 1¾ ounces (three ¼ + ⅛ + ¹⁄₁₆ color container) of the darkest brown tint to the developer.

5. Add ¼ ounce (⅛ color container) of soft black tint to the developer.

6. Shake the mixture until the materials are combined, holding a gloved finger over the open tip.

7. Apply the mixture from your scalp to the ends.

8. Comb through your hair to make sure the mixture is distributed evenly.

9. As soon as you are done with the application, saturate cotton balls with aerosol hair spray as needed and wipe away the triple-barrier cream in the direction of the hairline to avoid creating a harsh line. This ensures you will not stain your skin.

10. Let the color work for 20 minutes.

11. Remove the triple-barrier cream, rinse your hair, shampoo without manipulating the scalp. Condition and towel dry. Style as usual. Use Colorist Collection Cocoa Bean Shampoo every third shampoo to extend your color.

TOUCH-UP TIME

When you have regrowth, apply ½ the mixture to the regrowth area only and leave it in for 20 minutes. At 20 minutes, use a spray bottle to spritz water on the remaining hair. Add the remaining color to all of the hair and leave it in for 3 minutes. Then shampoo and condition your hair. Style as usual. (Or, try the foil-out technique for added texture!)

Tab 5

MATERIALS

Triple-barrier cream

Measuring cups and spoons

1 box dark brown

1 box black

8 cotton balls

Aerosol hair spray

Timer

Cleansing shampoo

L'Oréal Paris Colorist Collection Cocoa Bean Shampoo*

*Available at beauty supply stores and online (see the Resources section on page 289)

DIRECTIONS

1. Detangle your hair and comb it away from your face.

2. Apply the triple-barrier cream and put on gloves.

3. Snip off the tip from the applicator of the developer and remove the cap.

4. Using measuring spoons, add 1½ ounces plus ¼ ounce (three ¼ + ⅛ color container) of the dark brown tint to the developer.

5. Add ¼ ounce (⅛ color container) of the black tint to the developer.

6. Shake the mixture until the materials are combined, holding a gloved finger over the open tip.

7. Apply the mixture starting from your scalp to the ends. Comb through your hair so the mixture is distributed evenly. Leave it in for 5 minutes.

8. As soon as you are done with the application, saturate cotton balls with aerosol hair spray as needed and wipe away the triple-barrier cream in the direction of the hairline to avoid creating a harsh line. This ensures you will not stain your skin.

9. Let the color work for 25 minutes, until the timer goes off.

10. Rinse your hair, shampoo it, condition, and towel dry. Style as usual. Use Colorist Collection Cocoa Bean Shampoo every third shampoo to refresh your color.

Tab 5½

MATERIALS

Triple-barrier cream

Measuring cups and spoons

1 box dark brown

1 box black

8 cotton balls

Aerosol hair spray

Timer

Moisturizing shampoo

L'Oréal Paris Colorist Collection Cocoa Bean Shampoo*

Available at beauty supply stores and online (see the Resources section on page 289)

1. Detangle your hair and comb it away from your face.

2. Apply the triple-barrier cream and put on gloves.

3. Snip off the applicator tip of the developer and remove the cap.

4. Using measuring cups or spoons, add 1½ ounces (½ + ¼ color container) of the dark brown tint to the developer.

5. Add ½ ounce (¼ color container) of the black tint to the developer.

6. Shake the mixture until the materials are combined, holding a gloved finger over the open tip.

7. As soon as you are done with the application, saturate cotton balls with aerosol hair spray as needed and wipe away the triple-barrier cream in the direction of the hairline to avoid creating a harsh line. This ensures you will not stain your skin.

8. Apply mixture to the scalp to the ends. Comb the mixture through your hair so it is distributed evenly. Leave it on for 30 minutes.

9. Rinse your hair, shampoo, condition, and towel dry. Style as usual. Use Colorist Collection Cocoa Bean Shampoo every third shampoo to refresh your color.

Tab 6

MATERIALS

Triple-barrier cream

Measuring cups and spoons

1 box darkest brown

1 box black

1 box auburn brown or dark spice

8 cotton balls

Aerosol hair spray

Timer

Moisturizing shampoo

L'Oréal Paris Colorist Collection Cocoa Bean Shampoo*

Available at beauty supply stores and online (see the Resources section on page 289)

DIRECTIONS

1. Detangle your hair and comb it away from your face.

2. Apply the triple-barrier cream and put on gloves.

3. Snip off the applicator tip of the developer and remove the cap.

4. Using measuring cups or spoons, add ⅝ ounce (¼ + 1⁄16 color container) of the dark brown tint to the developer.

5. Add 1¼ ounces (½ + ⅛ color container) of the black tint to the developer.

6. Add ⅛ ounce (1⁄16 color container) of the auburn brown or dark spice tint to the developer.

7. Shake the mixture until the materials are combined, holding a gloved finger over the open tip.

8. Apply the mixture starting from your scalp to the ends. Comb through your hair so the mixture is distributed evenly.

9. As soon as you are done with the application, saturate cotton balls with aerosol hair spray as needed and wipe away the triple-barrier cream in the direction of the hairline to avoid creating a harsh line. This ensures you will not stain your skin.

10. Let the color work for 25 minutes.

11. Rinse your hair, shampoo, condition, and towel dry it. Style as usual. Use Colorist Collection Cocoa Bean Shampoo every third shampoo to refresh your color.

Tab 6½

If you have naturally light hair, I do not suggest that you go Espresso all over. You have to remember that this color is very tough to cover up or erase once it's been applied. Choosing to dye your hair Espresso is a step into high maintenance for a true light-haired girl. If you insist on Espresso, you should know that once you have ventured down this road and you want to make a change, you have a few options: do major corrective color in a salon, use your Color Oops (see Chapter 3), or wait for Espresso to grow out. As a fun alternative to all-over color, I recommend trying this recipe as a wide ribbon lowlight, also called a panel (the Peekaboo Punk technique). Apply it under the top layer of hair, next to the back of your ears or at your temples if you have longer hair that is not dramatically layered.

MATERIALS

Triple-barrier cream

Measuring cups and spoons

1 box darkest brown

1 box black

8 cotton balls

Aerosol hair spray

Timer

Moisturizing shampoo

L'Oréal Paris Colorist Collection Cocoa Bean Shampoo*

*Available at beauty supply stores and online (see the Resources section on page 289)

1. Detangle your hair and comb it away from your face.

2. Apply the triple-barrier cream and put on gloves.

3. Snip off the applicator tip of the developer and remove the cap.

4. Using measuring cups or spoons, add ¾ ounce (¼ + ⅛ color container) of the darkest brown tint to the developer.

5. Add 1½ ounces (½ + ⅛ color container) of the black tint to the developer. (If you want it darker next time, delete the dark brown by ⅛ ounce (1/16 color container) and increase the black by ⅛ ounce. If the color is still not deep enough for your liking, add the black in ⅛ ounce increments (1/16 color container) as you delete the dark brown tint in ⅛ ounce increments).

6. Shake the mixture until the materials are combined, holding a gloved finger over the open tip.

7. Apply the mixture as a wide ribbon highlight: Take an Oral-B toothbrush, dip it into the highlight, and paint or brush the highlight mixture starting from your scalp down to the ends. The top of the ribbon should be 1 to 2 inches wide. Each ribbon should be 1 to 2 inches from one another and staggered on either side of your part.

8. As soon as you are done with the application, saturate cotton balls with aerosol hair spray as needed and wipe away the triple-barrier cream in the direction of the hairline to avoid creating a harsh line. This ensures you will not stain your skin.

9. Set the timer for 30 minutes and relax!

10. Rinse your hair, shampoo, condition, and towel dry. Style as usual. Use Colorist Collection Cocoa Bean Shampoo every third shampoo to extend your color.

NOTE: If the finished result is not Espresso enough for you, next time add 1 ounce of dark brown and 1 ounce of black to the developer. Proceed as directed.

Please read the headnote for Tab 6 on page 70—it applies to these tabs as well.

MATERIALS

1 box light auburn

Triple-barrier cream

Measuring cups and spoons

1 box darkest brown

1 box black

Distilled water

8 cotton balls

Aerosol hair spray

Timer

Moisturizing shampoo

L'Oréal Paris Colorist Collection Cocoa Bean Shampoo*

*Available at beauty supply stores and online (see the Resources section on page 289)

DIRECTIONS

1. Apply the triple-barrier cream and put on gloves.

2. Snip off the applicator tip of the developer and remove the cap.

3. Using the measuring cups or spoons, add ¼ ounce (⅛ color container) of the darkest brown tint to the developer.

4. Add 1¾ ounces (three ¼ + ⅛ color container) of the black tint to the developer.

5. Shake the mixture until the materials are combined, holding a gloved finger over the open tip.

6. Apply the mixture starting from the scalp through to the ends.

7. Set the timer for 30 minutes.

8. As soon as you are done with the application, saturate cotton balls with aerosol hair spray as needed and wipe away the triple-barrier cream in the direction of the hairline to avoid creating a harsh line. This ensures you will not stain your skin.

9. Rinse your hair and shampoo it well. Condition and towel dry. Style as usual. Use Colorist Collection Cocoa Bean Shampoo every third shampoo to extend your color.

Tab 8

Tab 8s should follow the recipe for Tab 11. Just omit the pre-softening step.

Tab 9

Tab 9s should use the recipe for their pre-gray tab.

Tab 10

Tab 10s should revert to their pre-gray tab if they are closer to 35 percent white, or use the recipe for Tab 11 if they are closer to 65 percent white.

Tab 11

If you are a Tab 11, you must pre-soften first with a permanent solution.

MATERIALS

1 drugstore perm kit

1 box deep copper (such as L'Oréal Paris Féria)

1 box black (soft or natural)

Aerosol hair spray

Timer

8 cotton balls

Washcloth

L'Oréal Paris Colorist Collection Cocoa Bean Shampoo*

*Available at beauty supply stores and online (see the Resources section on page 289)

DIRECTIONS

1. Apply permanent solution from a drugstore perm kit to your hair for 3 minutes, then rinse lightly, shampoo with no scalp manipulation, and towel blot.

2. Pre-color with the deep copper.

3. Detangle your hair and comb it away from your face.

4. Apply the triple-barrier cream and put on gloves.

5. Snip off the applicator tip and remove the cap.

6. Pour the black tint into the developer and replace the cap.

7. Shake the mixture until combined, holding a gloved finger over the open tip.

8. Apply the mixture from the scalp through the ends.

9. As soon as you are done with the application, saturate cotton balls with aerosol hair spray and wipe away the triple-barrier cream in the direction of the hairline to avoid creating a harsh line.

10. Set the timer for 30 minutes and relax.

11. Shampoo your hair, then shampoo it again using Colorist Collection Cocoa Bean Shampoo. Rinse well, condition, and style as usual.

ESPRESSO WITH RASPBERRY LIGHTS

This recipe combines the rich, sultry depth of Espresso with luxe red highlights. The recipe for raspberry lights can be added after you have completed your tab recipe for Espresso. It's really a highlight with an added gloss. It has a custom salon look, but it is *so* simple to accomplish. If you already have espresso-colored hair naturally (Tab 2), apply this highlight formula to your hair to get the same result. Let's experience the sexy enhancement of raspberry lights!

Tab 1

MATERIALS

1 box lightest blonde

3 tablespoons cornstarch

Small wire whisk

Rattail comb

5–7 yoyettes

5–7 pieces of 3" x 6" heavy-duty foil

Plastic bowl

Oral-B soft toothbrush with an elongated end

1 can Dark and Lovely Color Flash 22, Raspberry Yum, or L'Oréal Paris Cherry Shine Mousse

Shampoo and conditioner

DIRECTIONS

1. Complete the Espresso recipe for Tab 1, part your hair in your favorite fashion, and dry your hair completely.

2. Mix the lightest blonde tint with the developer according to the package instructions and pour it into a plastic bowl.

3. Mix in the cornstarch using the small wire whisk. This thickens the mixture so it is easy to work with.

4. Using your rattail comb, pull ¼-inch sections of your hair (about 12 to 17 strands) along each side of your part, staggering the sections on each side. Clip the sections up with the yoyettes.

5. Next, take the Oral-B toothbrush, dip it into the blonde mixture, and brush it on yarn-size strands around your face starting around 1½ inches down from your scalp and keeping the application bottom heavy. (Refer to the balayage drawing on page 49.)

6. Set the timer for 10 minutes and wait.

7. Rinse and shampoo your hair gently with no scalp manipulation.

8. Towel blot and detangle your hair.

9. Apply the Color Flash 22, Raspberry Yum or Cherry Shine. I even use Cherry Shine in the salon. Shine extreme! Either way, apply the mousse according to the package directions, from your scalp to your ends.

10. Set the timer for 20 minutes.

11. When time is up, rinse your hair, shampoo gently with no scalp manipulation, and condition it. Style as usual.

NOTE: For even more intensity, I use L'Oréal Paris Colorist Collection Mahogany Shampoo every third cleansing for upkeep, along with a light conditioner for detangling.

MATERIALS

1 box of Clairol Frost'n Tip highlighting kit

Rattail comb

5–7 yoyettes

5–7 pieces of 3"x6" heavy-duty foil

Oral-B soft toothbrush with an elongated end

Water-jug paddle (see page 25)

Tung brush

1 can L'Oréal Paris Lively Auburn Mousse, or, if you are superdaring, 1 can L'Oréal Paris Cherry Shine Mousse

Shampoo and conditioner

L'Oréal Paris Colorist Collection Mahogany Shampoo*

Available at beauty supply stores and online (see the Resources section on page 289)

DIRECTIONS

1. Complete the Espresso recipe for Tab 2, 3, or 4 and dry your hair completely. (If hair is straight or permed, just do steps 2 to the end.)

2. Mix the Clairol Frost'n Tip kit according to the package directions.

3. Part your hair in your favorite way.

4. Using your rattail comb, pull ¼-inch sections of your hair (about 12 to 17 strands) along each side of your part, staggering sections on each side. Clip them up with the yoyettes.

5. Next, put on gloves. I take the Oral-B toothbrush, dip it into the blonde mixture, and brush it on yarn-size strands around your face, starting around 1½ inches down from your scalp and keeping the application bottom heavy. (Refer to the balayage

drawing on page 49.) Use the Tung brush to stroke the midsections of your hair. It does not have to be neat—we are adding texture, so perfection is not required. Be bold! This effect is subtle but needs a heavy hand during application.

6. Set the timer for 16 minutes. (For permed or straightened hair, start with minutes; if it's not intense enough, you can do 16 minutes next time.)

7. Rinse and shampoo your hair. Do not manipulate your scalp.

8. Towel blot and detangle your hair.

9. Apply L'Oréal Lively Auburn according to directions, starting from your scalp to your ends, or use L'Oréal Cherry Shine Mousse for a more intense color. I even use this in the salon. Shine extreme!

10. Leave the mousse in for 10 minutes.

11. Rinse your hair, shampoo gently with no scalp manipulation, condition, and towel dry. Style as usual.

12. To keep hair perfect, use Colorist Collection Mahogany Shampoo every third cleansing for upkeep. Implement your favorite mousse every two weeks on its own.

TOO MUCH OF A GOOD THING

If you apply too much highlighting mixture around your face, you can remove it with either olive or jojoba oil. Place the oil in a 1-ounce bottle and "blend" the oil on the excess highlighting mixture, then, if necessary, smooth it with a Q-tip.

Tab 5, 5½, and 6

It's important not to get too close to the scalp the first time you do this.

MATERIALS

1 box Clairol Frost'n Tip highlighting kit

1 box dark brown

Rattail comb

5–7 yoyettes

5–7 pieces of 3"x6" heavy-duty foil

Oral-B soft tip toothbrush with an elongated end

Tung brush

1 can L'Oréal Paris Lively Auburn Mousse

Shampoo and conditioner

L'Oréal Paris Colorist Collection Mahogany Shampoo*

Available at beauty supply stores and online (see the Resources section on page 289)

DIRECTIONS

1. Complete the Espresso recipe for Tabs 5, 5½, or 6 and dry your hair completely.

2. Mix the Clairol Frost'n Tip kit according to the package directions.

3. Part your hair in your favorite way.

4. Using your rattail comb, pull ¼-inch sections of hair (about 12 to 17 strands) along each side of your part, staggering sections on each side. Clip them up with the yoyettes.

5. Next, put on gloves. I take the Oral-B toothbrush, dip it into the blonde mixture, and brush it on yarn-size strands around your face starting around 1½ inches down from your scalp and keeping the application bottom heavy. (Refer to the balayage drawing on page 49.) Use the Tung brush to stroke the midsections of your hair. It does not have to be neat—we are adding texture, so perfection is not required. Be bold. This effect is subtle but needs a heavy hand during application.

6. Set the timer for 10 minutes.

7. Rinse and shampoo your hair gently with no scalp manipulation.

8. Towel blot and detangle your hair.

9. Apply L'Oréal Lively Auburn Mousse according to directions, starting from your scalp to your ends. (For brighter color, try L'Oréal Cherry Shine Mousse, but start at 5 minutes.)

10. Leave the mousse in for 10 minutes.

11. Rinse until the water runs clear and shampoo your hair. Condition and towel dry. Style as usual.

12. To keep the intensity, use Colorist Collection Mahogany Shampoo every third cleansing for upkeep. Implement your favorite color mousse on its own every two weeks.

Tabs 6½, 7, 7½, , 9, 10, and 11

Follow your tab recipe for Espresso, and then follow the highlight and gloss recipe for Tab 5½.

DARK CHOCOLATE

This is one chocolate craving you can definitely indulge in without any guilt!

Tab 1

MATERIALS

Triple-barrier cream

1 box medium dark golden brown (such as L'Oréal Paris Féria)

Clean washcloth

8 cotton balls

Aerosol hair spray

Timer

Cleansing shampoo and conditioner

L'Oréal Paris Colorist Collection Walnut Shampoo*

*Available at beauty supply stores and online (see the Resources section on page 289)

DIRECTIONS

1. Detangle your hair and comb it away from your face.

2. Apply the triple-barrier cream and put on gloves.

3. Snip off the applicator tip of the developer and remove the cap.

4. Add the dark golden brown tint to the developer and shake the mixture until the materials are combined, holding a gloved finger over the open tip.

5. Apply the mixture starting from 1½ inches down from your scalp to the ends, comb through your hair so the mixture is distributed, and leave it on for 5 minutes. Note that this is the smudge technique, page 50. Apply the mixture starting from your scalp to 1½ inches down, wait 3 minutes, and then shampoo.

6. For a touch-up application only, apply the color to the regrowth area, wait 3 minutes, and then shampoo.

7. Comb through your hair again to make sure the mixture is distributed evenly.

8. As soon as you are done with the application, saturate cotton balls with aerosol hair spray as needed and wipe away the triple-barrier cream in the direction of the hairline to avoid creating a harsh line. This ensures you will not stain your skin.

9. Immediately rinse your hair, shampoo vigorously with Colorist Collection Walnut, condition, and towel dry. Style as usual.

Tabs 2 and 3

MATERIALS

Triple-barrier cream

Measuring cups and spoons

1 box dark golden brown (such as L'Oréal Paris Féria)

1 box mahogany brown

Clean washcloth

8 cotton balls

Aerosol hair spray

Timer

Shampoo and conditioner

L'Oréal Paris Colorist Collection Walnut or Cocoa Bean Shampoo*

Available at beauty supply stores and online (see the Resources section on page 289)

DIRECTIONS

1. Detangle your hair and comb it away from your face.

2. Apply the triple-barrier cream and put on gloves.

3. Snip off the applicator tip of the developer and remove the cap.

4. Using measuring cups or spoons, add 1¾ ounces (three ¼ + ⅛ color container) of the dark golden brown tint to the developer.

5. Add ¼ ounce (⅛ color container) of the mahogany brown tint to the developer.

6. Shake the mix gently until the materials are combined, holding a gloved finger over the open tip.

7. Apply mixture 1½ inches down starting from your scalp to the ends, comb through your hair so the mixture is distributed evenly, and leave it on for 10 minutes.

8. Comb through your hair again to make sure the mixture is distributed evenly.

9. As soon as you are done with the application, saturate cotton balls with aerosol hair spray as needed and wipe away the triple-barrier cream in the direction of the hairline to avoid creating a harsh line. This ensures you will not stain your skin.

10. Rinse and shampoo your hair, then shampoo a second time with Colorist Collection Walnut. If you prefer cooler tones, use Colorist Collection Cocoa Bean. Condition and style as usual.

Tab 4

MATERIALS

Triple-barrier cream

Measuring cups and spoons

1 box dark golden brown

1 box mahogany brown

8 cotton balls

Aerosol hair spray

Timer

Shampoo and conditioner

L'Oréal Paris Colorist Collection Walnut or Cocoa Bean Shampoo*

*Available at beauty supply stores and online (see the Resources section on page 289)

1. Detangle your hair and comb it away from your face.

2. Apply the triple-barrier cream and put on gloves.

3. Snip off the applicator tip of the developer and remove the cap.

4. Using measuring cups or spoons, add 1¾ plus ⅛ ounces (three ¼ + ⅛ + ¹⁄₁₆ color container) of the dark golden brown tint to the developer.

5. Add ⅛ ounce (¹⁄₁₆ color container) of the mahogany brown tint to the developer.

6. Shake the mixture gently until the materials are combined, holding a gloved finger over the open tip.

7. Apply the mixture starting from your scalp to the ends and comb through your hair so the mixture is distributed evenly.

8. Set the timer for 30 minutes.

9. As soon as you are done with the application, saturate cotton balls with aerosol hair spray as needed and wipe away the triple-barrier cream in the direction of the hairline to avoid creating a harsh line. This ensures you will not stain your skin.

10. Rinse your hair and shampoo immediately. Shampoo a second time with Colorist Collection Walnut Shampoo, or try Colorist Collection Cocoa Bean if you like cooler tones. Condition and towel dry. Style as usual.

Tab 5

MATERIALS

Triple-barrier cream

Measuring cups and spoons

1 box dark golden brown

1 box dark brown

1 box mahogany brown

8 cotton balls

Aerosol hair spray

Timer

Shampoo and conditioner

L'Oréal Paris Colorist Collection Walnut or Cocoa Bean Shampoo*

*Available at beauty supply stores and online (see the Resources section on page 289)

DIRECTIONS

1. Detangle your hair and comb it away from your face.

2. Apply the triple-barrier cream and put on gloves.

3. Snip off the applicator tip of the developer and remove.

4. Using measuring cups or spoons, add 1 ounce (½ color container) of the dark golden brown tint to the developer.

5. Add 1 ounce (½ color container) of the dark brown tint to the developer.

6. Add ½ teaspoon of the mahogany brown tint to the developer.

7. Replace the developer cap and gently shake the mixture until the materials combine, holding a gloved finger over the open tip.

8. Apply the mixture starting from your scalp to the ends and comb through your hair so the mixture is distributed evenly.

9. As soon as you are done with the application, saturate cotton balls with aerosol hair spray as needed and wipe away the triple-barrier cream in the direction of the hairline to avoid creating a harsh line. This ensures you will not stain your skin.

10. Let the mixture work for 30 minutes.

11. When time is up, rinse your hair, shampoo well, do a second shampoo with Colorist Collection Walnut or Cocoa Bean Shampoo, condition, and towel dry. Style as usual.

NOTE: If you want a warmer result, next time add an additional ¼ teaspoon of the mahogany brown tint. No additional developer is necessary. If you want it to be darker, next time increase the dark brown by ¼ teaspoon and delete the dark golden brown tint. Like any real recipe, some people like a bit more cinnamon or nutmeg, and these changes can give you that extra "flavor."

Tabs 5½ and 6

MATERIALS

Triple-barrier cream

Measuring cups and spoons

1 box dark brown

1 box dark golden brown

1 box mahogany

8 cotton balls

Aerosol hair spray

Timer

Shampoo and conditioner

L'Oréal Paris Colorist Collection Walnut or Cocoa Bean Shampoo*

Available at beauty supply stores and online (see the Resources section on page 289)

DIRECTIONS

1. Detangle your hair and comb it away from your face.

2. Apply the triple-barrier cream and put on gloves.

3. Snip off the applicator tip of developer and remove the cap.

4. Using measuring spoons or cups, add 1½ ounces (three ¼ color container) of the dark brown tint to the developer.

5. Add ½ ounce (¼ color container) of the dark golden brown tint to the developer.

6. Add ½ teaspoon of the mahogany brown tint to the developer.

7. Replace the developer cap and shake the mixture until the materials are combined, holding a gloved finger over the open tip.

8. Apply the mixture starting from your scalp to the ends and comb through your hair.

9. Comb through your hair again to make sure the mixture is distributed evenly.

10. As soon as you are done with the application, saturate cotton balls with aerosol hair spray as needed and wipe away the triple-barrier cream in the direction of the hairline to avoid creating a harsh line. This ensures you will not stain your skin.

11. Let the color work for 30 minutes.

12. When the time is up, rinse your hair, shampoo, shampoo again using Colorist Collection Walnut Shampoo (or Colorist Collection in Cocoa Bean, if you like cooler tones), condition, and towel dry. Style as usual.

Just a reminder to naturally light-haired girls—this is stepping into high-maintenance color. If you don't like it, you'll have to do a major corrective color in a salon, use Color Oops (see Chapter 3), or wait for it to grow out. As a fun alternative, I recommend trying this recipe as a wide ribbon panel (see the Peekaboo Punk technique, page 236).

Tab 6½

MATERIALS

Triple-barrier cream

Measuring cups and spoons

1 box dark brown

1 box medium red brown

1 box medium auburn

8 cotton balls

Aerosol hair spray

Timer

Shampoo and conditioner

L'Oréal Paris Colorist Collection Walnut or Cocoa Bean Shampoo*

Available at beauty supply stores and online (see the Resources section on page 289)

DIRECTIONS

1. Detangle your hair and comb it away from your face.

2. Apply the triple-barrier cream and put on gloves.

3. Snip off the applicator tip of the developer and remove the cap.

4. Using measuring cups or spoons, add 1 ounce (½ color container) of the dark brown tint to the developer.

5. Add 1 ounce (½ color container) of the medium red brown tint to the developer.

6. Add 1 teaspoon medium auburn.

7. Replace the developer cap and gently shake the mixture until the materials are combined, holding a gloved finger over the open tip.

8. Apply the mixture starting from your scalp through to the ends.

9. Comb the mixture through your hair to make sure it is distributed evenly.

10. As soon as you are done with the application, saturate cotton balls with aerosol hair spray as needed and wipe away the triple-barrier cream in the direction of the hairline to avoid creating a harsh line. This ensures you will not stain your skin.

11. Let the color work for 30 minutes.

12. When time is up, remove the triple-barrier cream with a clean washcloth, rinse your hair, shampoo well, shampoo again using Colorist Collection Walnut Shampoo (or Colorist Collection in Cocoa Bean, if you prefer a cooler tone), condition, and towel dry. Style as usual.

Tabs 7 and 7½

MATERIALS

Triple-barrier cream

Measuring cups and spoons

1 box brown black

1 box dark golden brown

1 box mahogany brown

8 cotton balls

Aerosol hair spray

Timer

Shampoo and conditioner

L'Oréal Paris Colorist Collection Walnut Shampoo*

Available at beauty supply stores and online (see the Resources section on page 289)

DIRECTIONS

1. Detangle your hair and comb it away from your face.

2. Apply the triple-barrier cream and put on gloves.

3. Snip off the applicator tip of the developer and remove the cap.

4. Using measuring cups or spoons, and add 1¼ ounces (½ + ⅛ color container) of the brown black tint to the developer.

5. Add ½ ounce (¼ color container) of the dark golden brown tint to the developer.

6. Add ¼ ounce (⅛ color container) of the mahogany brown tint to the developer.

7. Replace the developer cap and shake the mixture until the materials are combined, holding a gloved finger over the open tip.

8. Apply the mixture from your scalp to the ends and comb through your hair so the mixture is distributed evenly.

9. Comb through your hair again to make sure the mixture is distributed evenly.

10. As soon as you are done with the application, saturate cotton balls with aerosol hair spray as needed and wipe away the triple-barrier cream in the direction of the hairline to avoid creating a harsh line. This ensures you will not stain your skin.

11. Let the color work for 20 minutes.

12. When time is up, rinse your hair, shampoo well, shampoo again with Colorist Collection in Walnut, condition, and towel dry. Style as usual. Use Colorist Collection in Walnut for every third shampoo to help maintain your color.

Tab 8

MATERIALS

Triple-barrier cream

Measuring cups and spoons

1 box brown black

1 box mahogany brown

8 cotton balls

Aerosol hair spray

Timer

Shampoo and conditioner

L'Oréal Paris Colorist Collection Walnut Shampoo*

Available at beauty supply stores and online (see the Resources section on page 289)

DIRECTIONS

1. Detangle your hair and comb it away from your face.

2. Apply the triple-barrier cream and put on gloves.

3. Snip off the applicator tip of the developer and remove the cap.

4. Using measuring cups or spoons, 1¾ ounces (three ¼ + ⅛ color container) of brown black tint to the developer.

5. Add ¼ ounce (⅛ color container) of mahogany brown tint to the developer.

6. Replace the developer cap and gently shake the mixture until the materials are combined, holding a gloved finger over the open tip.

7. Apply the mixture starting from your scalp to the ends and comb through your hair so the mixture is distributed evenly.

8. Comb through your hair again to make sure the mixture is distributed evenly.

9. As soon as you are done with the application, saturate cotton balls with aerosol hair spray as needed and wipe away the triple-barrier cream in the direction of the hairline to avoid creating a harsh line. This ensures you will not stain your skin.

10. Let the color work for 30 minutes.

11. When time is up, rinse your hair, shampoo, shampoo again with Colorist Collection Walnut Shampoo, condition, and towel dry. Style as usual. Use Colorist Collection Walnut for every third shampoo to help maintain your color.

Tab 9

Follow the recipe for your natural pre-gray tab. If it's still not deep enough for you, next time add ⅛ ounce (¹⁄₁₆ color container) of the darkest tint and remove ⅛ ounce of the lightest tint.

Tab 10

If you are closer to 35 percent gray, follow your pre-gray tab; if you are closer to 65 percent gray, follow the Tab 11 recipe.

Tab 11

MATERIALS

1 perm kit

Triple-barrier cream

Measuring cups and spoons

1 box dark brown (such as Revlon Colorist 40 or Clairol Nice'n Easy 120)

1 box medium red brown (such as Revlon Colorist 57 or Clairol Nice'n Easy 130)

1 box medium auburn (such as Revlon Colorist 55 or Clairol Nice'n Easy 111)

8 cotton balls

Gray Magic*

Aerosol hair spray

Timer

Shampoo and conditioner

L'Oréal Paris Colorist Collection Walnut Shampoo*

*Available at beauty supply stores and online (see the Resources section on page 289)

DIRECTIONS

1. Pre-soften hair: Apply the perm solution to your hair starting about 1¼ inches from the scalp down to the ends and let it work for 3 minutes. Rinse, shampoo with no scalp manipulation, lightly comb out your hair, and towel blot. You are ready to continue with your recipe.

2. Apply the triple-barrier cream and put on gloves.

3. Snip off the applicator tip of the developer and remove the cap.

4. Using measuring cups or spoons, add 1¼ ounces (½ + ⅛ color container) of the dark brown tint to the developer.

5. Add ¾ ounce (¼ + ⅛ color container) of medium red brown tint to the developer.

6. Add 1 teaspoon medium auburn tint to the developer.

7. If desired, add 2 drops of Gray Magic to get better coverage.

8. Replace the developer cap and shake the mixture until the materials are combined, holding a gloved finger over the open tip.

9. Apply the mixture starting from your scalp to the ends, comb through your hair so the mixture is distributed evenly, and leave it on for 30 minutes.

10. As soon as you are done with the application, saturate cotton balls with aerosol hair spray as needed and wipe away the triple-barrier cream in the direction of the hairline to avoid creating a harsh line. This ensures you will not stain your skin.

11. When time is up, rinse your hair, shampoo well, and shampoo a second time with Colorist Collection Walnut Shampoo or one of the redder Revlon color glazes. Condition, towel dry, and style as usual. If you prefer a cooler or less red Dark Chocolate, use Colorist Collection Cocoa Bean Shampoo or Revlon Colorist Dark Brown Color Glaze. Use this every third shampoo to extend your color.

DARK CHOCOLATE WITH DEEP CARAMEL RIBBONS

After you complete your Dark Chocolate recipe, you may want to add some highlights. These deep caramel ribbons are just the thing for rich, sexy hair. Doesn't deep caramel *always* bring chocolate to a whole new level? If you have naturally dark hair, use the highlight recipe to add some deep caramel to your color.

Tabs 1 through 10

MATERIALS

Old button-down shirt

1 box Clairol Frost'n Tip highlighting kit

40-volume peroxide*

Newspaper or paper or canvas drop cloth for floor work area

1 teaspoon (approximately) Vaseline petroleum jelly

10 to 20 cotton balls

2 Oral-B soft toothbrushes with elongated tips

1 or 2 teaspoons olive or vegetable oil (this helps the texture, because at-home kits are so dry)

2 to 4 white washcloths

Plastic wrap

1 plastic bowl (If a container is provided in the kit, use that)

1 hair dryer with diffuser attached or your cotton sock diffuser

2 yoyettes

Tung brush

Timer

Shampoo and conditioner

L'Oréal Paris Colorist Collection Walnut Shampoo*

Available at beauty supply stores and online (see the Resources section on page 289)

1. Part your hair in your favorite place and clip flat with yoyettes.

2. Make sure your hair is completely dry after completing your tab recipe for Dark Chocolate. Put on gloves (I prefer to use recycled ones from the color kits).

3. Open the cream peroxide in the Frost'n Tip highlighting kit and take the cap off the 40-volume beauty supply peroxide.

4. Using the cap off of the 40-volume peroxide as a measure, pour two caps of the Frost'n Tip peroxide and discard. Rinse the cap, and, using it as a measure, pour 2 caps of the 40-volume peroxide into the Frost'n Tip peroxide bottle. You have simply replaced some of the Frost'n Tip peroxide with a more powerful 40-volume peroxide.

5. Begin the standard highlight procedure. Prepare your work area, place the drop cloth under your feet, and put on your button-down work shirt.

6. Spread petroleum jelly on 3 unrolled cotton balls and stick them to your forehead under your hairline and the sides of your face in front of your hairline.

7. Mix the bleach, developer, and oil in the bowl.

8. Starting in front of your ear, separate a ¼-inch ribbon and clip it up with a yoyette. Take another section of the same size 1 inch or so away from the last section and clip it up, working that way on either side of your head. (See the balayage technique on page 49).

9. When you are done, one at a time, unclip each section, hold each one very taut, dip the tip of one of the Oral-B toothbrushes into the bleach mix, and coat the ribbon by applying the mixture starting from your scalp to mid hair shaft. Then dip the Tung brush into the mixture and brush it through to the ends. (If hair is long, brush on top of a paddle.)

10. Wipe your gloves on a washcloth after each stroke is finished. Try to make the mixture finer at the scalp and thicker through the ends.

11. When you finish, lay cotton strips over the highlights, and then sheets of plastic wrap, and start timing. Set the timer for 10 minutes.

12. Spot heat each section of your hair for 1 minute (with dryer and diffuser attached). Always do your ends first as they have no scalp heat to help them along. I personally use this technique in the salon and on my house calls, and the results are gorgeous!

13. Check the ribbons to see if their color matches dark caramel candy. If not, leave the mixture in 5 to 7 minutes longer, unless your hair is really resistant. Add 5-minute increments until the highlights match dark caramel.

14. Rinse your hair and shampoo with a clear color moisturizing shampoo mixture, such as Pantene PRO-V with a dime-size dollop of Colorist Collection Walnut. Use the Colorist Collection mixture every third shampoo.

Tab 11

MATERIALS

1 perm kit

Triple-barrier cream

Measuring cups and spoons

1 box dark brown (such as Revlon Colorist 40 or Clairol Nice'n Easy 120)

1 box medium red brown (such as Revlon Colorist 57 or Clairol Nice'n Easy 130)

1 box medium auburn (such as Revlon Colorist 55 or Clairol Nice'n Easy 111)

8 cotton balls

Gray Magic*

Aerosol hair spray

Timer

Shampoo and conditioner

L'Oréal Paris Colorist Collection Walnut Shampoo*

*Available at beauty supply stores and online (see the Resources section on page 289)

1. Pre-soften hair: Apply the perm solution to your hair starting about 1¼ inches from the scalp down to the ends and let it work for 3 minutes. Rinse, shampoo with no scalp manipulation, lightly comb out your hair, and towel blot. You are ready to continue with your recipe.

2. Apply the triple-barrier cream and put on gloves.

3. Snip off the applicator tip of the developer and remove the cap.

4. Using measuring cups or spoons, add 1¼ ounces (½ + ⅛ color container) of the dark brown tint to the developer.

5. Add ¾ ounce (¼ + ⅛ color container) of medium red brown tint to the developer.

6. Add 1 teaspoon medium auburn tint to the developer.

7. If desired, add 2 drops of Gray Magic to get better coverage.

8. Shake the mixture until the materials are combined, holding a gloved finger over the open tip.

9. Apply the mixture starting from your scalp to the ends, comb through your hair so the mixture is distributed evenly, and leave it on for 30 minutes.

10. As soon as you are done with the application, saturate cotton balls with aerosol hair spray as needed and wipe away the triple-barrier cream in the direction of the hairline to avoid creating a harsh line. This ensures you will not stain your skin.

11. When time is up, rinse your hair, shampoo well, and shampoo a second time with Colorist Collection Walnut Shampoo or one of the redder Revlon color glazes. Condition, towel dry, and style as usual. If you prefer a cooler or less red Dark Chocolate, use Colorist Collection Cocoa Bean Shampoo or Revlon Colorist Dark Brown Color Glaze. Use this every third shampoo to extend your color.

5.

RICH AND SUMPTUOUS BRUNETTES

T his chapter will show you how to add warmth and light to your favorite milk-chocolate hue. The colors are luxurious and feminine. The target colors are:

- **MILK CHOCOLATE**

- **CREAM SWIRLED MILK CHOCOLATE**

- **CHERRY COLA**

NOTE: If you have very thick, long hair (shoulder length or longer), double all recipes.

NOTE: All recipes give instructions for virgin hair. For instructions for retouching or any other technique, refer to Chapter 3.

NOTE: If you have very short hair, use the Frosted Secret Sauce highlight technique found in Chapter 10, on men's hair color, page 247.

MILK CHOCOLATE

Tab 1

MATERIALS

Triple-barrier cream

Timer

L'Oréal Paris Féria High Lift Brown #B51

Shampoo and conditioner

L'Oréal Paris Colorist Collection Walnut Shampoo*

*Available at beauty supply stores and online (see the Resources section on page 289)

READ ME! PREPARATION IS EVERYTHING

Before you begin any recipe, read through the ingredient list, get out your color kit, and prepare your work area. Clear off the countertop on which you will be working, and spread out your materials. For all-over colors, dampen your hair and towel blot. Comb out your hair. Put on your work shirt and wrap a towel around your neck to catch stains. Apply the triple-barrier cream around your hairline (page 37). This is especially important for brunette recipes. Cover the floor with a drop cloth or old newspapers. (Remember, no plastic underfoot!) Be sure to have your timer out and ready to set. *Wear kit-provided gloves when applying color!* You can recycle the gloves by rinsing them inside and out, patting them dry, and sprinkling some baby powder or cornstarch on the inside of them, for easy slip-on next time.

1. Detangle your hair and comb it away from your face.

2. Apply the triple-barrier cream and put on gloves.

3. Prepare L'Oréal Féria High Lift Brown #B51 and apply it starting from the mid-shaft of your hair to the ends and wait 5 minutes. Then apply it from your scalp to the mid-shaft of your hair and comb through your hair.

4. Set the timer for 5 minutes.

5. When time is up, rinse and shampoo your hair, removing the triple-barrier cream as you wash, then condition and towel dry your hair. Style as usual. Use Colorist Collection in Walnut Shampoo every third shampoo to maintain your color.

Tabs 2, 3, and 4

If you're doing a touch-up, start with step 4.

MATERIALS

Triple-barrier cream

1 box light golden blonde

1 box light golden brown

8 cotton balls

Aerosol hair spray

Timer

Shampoo and conditioner

L'Oréal Paris Colorist Collection Walnut Shampoo*

Available at beauty supply stores and online (see the Resources section on page 289)

1. Mix the light golden blonde tint according to the package instructions and apply it midway down your hair to the ends.

2. Set the timer for 5 minutes.

3. When time is up, shampoo your hair gently with no scalp manipulation and towel blot it.

4. Apply the triple-barrier cream and put on gloves.

5. Prepare the light golden brown according to the package instructions and apply it starting from your scalp all the way through to the ends. (For touch-ups, apply to regrowth only.)

6. As soon as you are done with the application, saturate cotton balls with aerosol hair spray as needed and wipe away the triple-barrier cream in the direction of the hairline to avoid creating a harsh line. This ensures you will not stain your skin.

7. Set the timer for 10 minutes.

8. When time is up, rinse your hair, and shampoo. Do a second shampoo with a mixture of half moisturizing shampoo and half Colorist Collection Walnut Shampoo. Rinse well, condition, and towel dry. Style as usual.

TRIPLE-BARRIER CREAM—A REMINDER

First, with your fingertips, apply a layer of lotion, such as Nivea or Keri brand, around your face at your hairline. Follow that with a layer of petroleum jelly. Then along the outside border of the lotion/petroleum jelly strip, apply a layer of Carmex lip balm.

Tab 5

MATERIALS

Triple-barrier cream

2 boxes light golden brown

8 cotton balls

Aerosol hair spray

Timer

Shampoo and conditioner

L'Oréal Paris Colorist Collection Walnut Shampoo*

Available at beauty supply stores and online (see the Resources section on page 289)

DIRECTIONS

1. Detangle your hair and comb it away from your face. Put on gloves.

2. Mix 1 box light golden brown according to the package instructions and apply the mixture starting from midway down your hair to the ends.

3. Set the timer for 5 minutes and apply the triple-barrier cream.

4. When time is up, prepare the second box of light golden brown according to the package instructions and apply it starting from your scalp all the way through to the ends.

5. As soon as you are done with the application, saturate cotton balls with aerosol hair spray as needed and wipe away the triple-barrier cream in the direction of the hairline to avoid creating a harsh line. This ensures you will not stain your skin.

6. Set the timer for 7 minutes.

7. When time is up, immediately rinse your hair, and shampoo. Do a second shampoo with a mixture of half moisturizing shampoo and half Colorist Collection Walnut Shampoo. Rinse well, condition, and towel dry. Style as usual.

MATERIALS

Triple-barrier cream

1 box medium golden brown

8 cotton balls

Aerosol hair spray

Timer

Shampoo and conditioner

L'Oréal Paris Colorist Collection Walnut Shampoo*

Available at beauty supply stores and online (see the Resources section on page 289)

DIRECTIONS

1. Detangle your hair and comb it away from your face.

2. Apply the triple-barrier cream and put on gloves.

3. Prepare the medium golden brown according to the package directions and apply it starting from your scalp to the ends.

4. As soon as you are done with the application, saturate cotton balls with aerosol hair spray as needed and wipe away the triple-barrier cream in the direction of the hairline to avoid creating a harsh line. This ensures you will not stain your skin.

5. Set the timer for 30 minutes.

6. When time is up, rinse your hair and shampoo. Shampoo again with a mixture of half moisturizing shampoo and half Colorist Collection in Walnut. Rinse well, condition, and towel dry. Style as usual. Use the Colorist Collection mixture for every third shampoo to help extend your color.

A PRECAUTIONARY NOTE FOR ALL YOU FAIR-HAIRED LASSES: Remember that this color is very tough to cover up or erase once it has been applied. You will have to do a major corrective color in a salon, use Color Oops (see

Chapter 3), or wait for Milk Chocolate to grow out. As an alternative, I prefer this recipe color as a lowlight using the Peekaboo Punk technique (see Chapter 236).

Tab 6½

Triple-barrier cream

Measuring cups and spoons

1 box medium golden brown

1 box brown

8 cotton balls

Aerosol hair spray

Timer

Shampoo and conditioner

L'Oréal Paris Colorist Collection Walnut Shampoo*

Available at beauty supply stores and online (see the Resources section on page 289)

DIRECTIONS

1. Detangle your hair and comb it away from your face.

2. Apply the triple-barrier cream and put on gloves.

3. Snip off the developer application tip, and remove the cap.

4. Using measuring cups or spoons, add 1¾ ounces (three ¼ + ⅛ color container) of medium golden brown tint to the developer.

5. Add ¼ ounce (⅛ color container) of the brown tint to the developer.

6. Shake the developer bottle gently to mix the tints and the developer completely, placing your gloved finger over the open tip.

7. Apply the mixture starting from your scalp down through to your ends. Comb through your hair to make sure the mixture is distributed evenly.

8. As soon as you are done with the application, saturate cotton balls with aerosol hair spray as needed and wipe away the triple-barrier cream in the direction of the hairline to avoid creating a harsh line. This ensures you will not stain your skin.

9. Set the timer for 30 minutes.

10. When time is up, rinse your hair, shampoo with no scalp manipulation, and towel dry. Style as usual.

Tabs 7 and 7½

MATERIALS

Timer

1 box dark golden brown

1 box medium golden brown

8 cotton balls

Aerosol hair spray

Shampoo and conditioner

L'Oréal Paris Colorist Collection Walnut Shampoo*

*Available at beauty supply stores and online (see the Resources section on page 289)

DIRECTIONS

1. Detangle your hair and comb it away from your face.

2. Snip the tip off the developer bottle and remove the cap. Put on gloves.

3. Using measuring cups or spoons, add 1½ ounces (three ¼ color container) of dark golden brown to the developer bottle.

4. Add ½ ounce (¼ color container) of medium golden brown.

5. Shake the bottle to combine the tints and developer, keeping your gloved finger over the open tip.

6. Apply the mixture to your hair, starting from your scalp down through the ends.

7. Set the timer for 30 minutes.

8. When time is up, rinse your hair, shampoo well, and use Colorist Collection Walnut Shampoo as a second shampoo. Rinse well, lightly condition, towel dry, and style as usual.

Tab 8

MATERIALS

Timer

1 box medium golden brown

1 box dark golden brown

8 cotton balls

Aerosol hair spray

Shampoo and conditioner

L'Oréal Paris Colorist Collection Walnut Shampoo*

Available at beauty supply stores and online (see the Resources section on page 289)

DIRECTIONS

1. Detangle your hair and comb it away from your face.

2. Snip the tip off the developer bottle and remove the cap. Put on gloves.

3. Using measuring cups or spoons, add 1 ounce (½ color container) of dark golden brown to the developer bottle.

4. Add 1 ounce (½ color container) medium golden brown.

5. Shake the bottle to combine the tints and the developer, keeping your gloved finger over the open tip.

6. Apply the mixture to your hair, starting from your scalp down through to the ends.

7. Set the timer for 30 minutes.

8. When time is up, rinse your hair, shampoo well, and use Colorist Collection Walnut Shampoo as a second shampoo. Rinse well, lightly condition, and towel dry. Style as usual.

Tab 9

Revert to the recipe for your pre-gray natural color tab.

Tab 10

Revert to the recipe for your pre-gray natural tab color. If your hair is closer to 35 percent gray, use the recipe for Tabs 5½ and 6. If your hair is closer to 65 percent gray, use the Tab 11 recipe.

Tab 11

MATERIALS

Perm kit

Triple-barrier cream

1 box dark brown

1 box medium red or reddish brown

1 box auburn

8 cotton balls

Aerosol hair spray

Measuring cups and spoons

Timer

Shampoo and conditioner

L'Oréal Paris Colorist Collection Walnut Shampoo*

Available at beauty supply stores and online (see the Resources section on page 289)

1. Pre-soften hair: Apply the permanent solution from a drugstore perm kit to your hair for 3 minutes, then rinse off and towel dry. Simply apply the solution to your hair starting about 1¼ inches from your scalp down to the ends, and let it work for 1 minute. Rinse, shampoo, and comb out your hair. You are ready to continue with your recipe.

2. Snip off the tip of the developer bottle in the box of dark brown.

3. Using measuring cups or spoons, add 1 ounce (½ color container) of dark brown.

4. Add 1 ounce (½ color container) of medium red or reddish brown.

5. Add 1 teaspoon of the auburn.

6. Shake the developer bottle to combine the mixture, keeping your gloved finger over the open tip.

7. Apply the mixture to your hair, starting from your scalp through to the ends, and then comb through your hair to make sure the mixture is distributed evenly.

8. As soon as you are done with the application, saturate cotton balls with aerosol hair spray as needed and wipe away the triple-barrier cream in the direction of the hairline to avoid creating a harsh line. This ensures you will not stain your skin.

9. Set the timer for 30 minutes.

10. When time is up, rinse your hair and shampoo. Shampoo again with a mixture of half moisturizing shampoo and half Colorist Collection in Walnut. Rinse well, condition lightly, and towel dry. Style as usual.

CREAM-SWIRLED MILK CHOCOLATE

After you have completed the Milk Chocolate process and your hair is completely dry, you can add some delicious swirls of cream all around your face. It is a two-step process: First you apply highlights, and then you add a finishing gloss on top of damp, detangled, towel-blotted hair. If you have naturally dark hair, use this highlight formula alone to add some cream to your coffee (or chocolate)!

Tab 1

MATERIALS

Rattail comb

Plastic bowl

1½ teaspoons olive oil

1 box Clairol Frost'n Tip Dramatic Highlight Bleach

40-volume peroxide

Oral-B soft toothbrush with an elongated end

Tung brush

10 to 12 unrolled cotton balls

Plastic wrap

Timer

Hair dryer with diffuser or diffuser sock

Shampoo and conditioner

1 box light golden brown

Measuring cups and spoons

Distilled water

L'Oréal Paris Colorist Collection Cocoa Bean Shampoo*

*Available at beauty supply stores and online (see the Resources section on page 289)

1. Complete Tab 1 for the Milk Chocolate recipe and dry hair completely. Once your hair is tinted, you must use the bleach mixture to achieve the creamy swirl effect.

2. Part your hair in your favorite way.

3. In a plastic bowl, combine the olive oil, Frost'n Tip Bleach, and 2 caps of the 40-volume peroxide. Rinse the cap and use it to remove and discard 2 caps of the enclosed developer. Add the remaining developer to the bowl and mix well.

4. Next, take the Oral-B toothbrush, dip it into the blonde mixture, and brush it on yarn-size strands around your face starting around 1½ inches down from your scalp and keeping the application bottom heavy. It does not have to be neat—we are adding texture, so perfection is not required. Dip the Tung brush into the mixture and brush it through to the ends. Be bold! (See the balayage technique on page 49.)

5. Cover your hair with unrolled cotton balls and plastic wrap.

6. Set the timer for 5 minutes, then hover with your hair dryer and diffuser, concentrating on the ends. You're looking for the highlight to match the color of a lemon; check when the 5 minutes is done. Use the hair dryer for no more than 25 minutes.

7. Rinse and shampoo your hair with no scalp manipulation. Towel blot and detangle.

8. Take out the box of light golden brown. Snip off the tip of the developer bottle and remove the cap.

9. Using measuring cups or spoons, remove 1 ounce (½ developer container) of developer. Add in 1 ounce of distilled water. Add the tint to the developer and shake the bottle until mixed thoroughly, placing your gloved finger over the open tip.

10. Apply the mixture to your hair starting from your scalp down to the ends and comb through your hair to make sure the mixture is distributed evenly.

11. Set the timer for 5 minutes.

12. When time is up, rinse your hair, shampoo gently with no scalp manipulation, and condition lightly. Towel dry and style as usual. Use Colorist Collection Cocoa Bean Shampoo every third shampoo to maintain your color.

Tabs 2, 3, and 4

MATERIALS

1 box Clairol Frost'n Tip highlighting kit

Rattail comb

5–7 yoyettes

Oral-B soft toothbrush with an elongated end

1 box light golden brown

Measuring cups and spoons

Distilled water

Timer

Shampoo and conditioner

DIRECTIONS

1. Complete the Milk Chocolate recipe for Tab 2, 3, or 4 and dry hair completely.

2. Mix the Clairol Frost'n Tip kit according to the package directions.

3. Part your hair in your favorite way.

4. Using your rattail comb, pull ¼-inch sections of hair (about 12 to 17 strands) along each side of your part, staggering sections on each side. Clip them up with yoyettes.

5. Dip the Oral B toothbrush into the bleach mixture and brush it on each strand, starting about 1½ inches down from your scalp and keeping the application bottom heavy.

6. Next, take the Tung brush, dip it into the blonde mixture, and brush it on yarn-size strands around your face starting 1½ inches down from your scalp and keeping the application bottom heavy. It does not have to be neat—we are adding texture, so perfection is not required.

7. Put the unrolled cotton balls over the highlights and drape plastic wrap on top.

8. Hover with your hair dryer and diffuser over the ends for counts of 20.

9. Look at the highlights to see if they match the color of a lemon. Set the timer so that you check the highlights every 5 minutes until you reach the desired result, for a total of no more than 30 minutes.

10. Rinse and shampoo your hair with no scalp manipulation.

11. Towel blot and detangle hair.

12. Take out the box of light golden brown. Snip off the tip of the developer bottle and remove the cap.

13. Using measuring cups or spoons, remove 1 ounce of developer (½ developer container). Add 1 ounce distilled water. Add the tint to the developer and shake the bottle until mixed thoroughly, placing your gloved finger over the open tip.

14. Apply the mixture to your hair starting from your scalp down to the ends and comb through your hair to make sure it is evenly distributed.

15. Set the timer for 5 minutes.

16. When time is up, rinse your hair, shampoo gently with no scalp manipulation, and condition lightly. Towel dry and style as usual.

Tabs 5, 5½, and 6

MATERIALS

1 box Clairol Frost'n Tip highlighting kit

Plastic spoon

Rattail comb

5–7 yoyettes

Oral-B soft toothbrush with an elongated end

1 box light golden brown

Measuring cups and spoons

Distilled water

Timer

Shampoo and conditioner

DIRECTIONS

1. Complete the Milk Chocolate recipe for Tab 5, 5½, or 6 and dry hair completely.

2. Part your hair in your favorite way.

3. Using your rattail comb, pull ¼-inch sections of hair (about 12 to 17 strands) along each side of your part, staggering sections on each side. Clip them up with yoyettes.

4. Mix the Clairol Frost'n Tip kit according to the package directions.

5. Dip the Oral B toothbrush into the bleach mixture and brush it on each strand, starting about 1½ inches down from your scalp and keeping the application bottom heavy.

6. Next, take the Tung brush, dip it into the blonde mixture, and brush it on yarn-size strands around your face starting 1½ inches down from your scalp and keeping the application bottom heavy. It does not have to be neat—we are adding texture, so perfection is not required.

7. Put the unrolled cotton balls over the highlights and drape plastic wrap on top.

8. Hover with your hair dryer and diffuser over the ends for counts of 20.

9. Look at the highlights to see if they match the color of a lemon. Set the timer so that you check the highlights every 5 minutes until you reach the desired result, for a total of no more than 30 minutes.

10. Rinse your hair and shampoo gently with no scalp manipulation. Towel blot and detangle your hair.

11. Take out the box of light golden brown. Snip off the tip of the developer bottle and remove the cap.

12. Using measuring cups or spoons, remove 1 ounce of developer (½ developer container). Add 1 ounce of distilled water. Add the tint to the developer and shake the bottle until mixed thoroughly, placing your gloved finger over the open tip.

13. Apply the mixture to your hair starting from your scalp down to the ends, and comb through your hair to make sure it is distributed evenly.

14. Set the timer for 5 minutes.

15. When time is up, rinse your hair, shampoo gently with no scalp manipulation, and condition lightly. Towel dry and style as usual.

Tab 6½

If you must, complete the Milk Chocolate recipe for Tab 6½ and then follow the highlight recipe for Tab 5½.

Tabs 7, 7½, 8, 9, 10, and 11

Follow your tab recipe for Milk Chocolate, follow the highlight recipe for Tabs 5, 5½, and 6, and then add the gloss recipe for Tab 5½.

CHERRY COLA

This effervescent all-over color will make any brunette bubble up with enthusiasm—so why not serve up a glass of this fabulous shade?

Tab 1

MATERIALS

1 box light blonde

Plastic bowl

Tung brush

Homemade water-jug paddle

8 unrolled cotton balls

Plastic wrap

Timer

Hair dryer with diffuser or diffuser sock

Shampoo and conditioner

Triple-barrier cream

1 box auburn brown (L'Oréal Paris Féria #56 is great)

Aerosol hair spray

L'Oréal Paris Colorist Collection Mahogany Shampoo*

Available at beauty supply stores and online (see the Resources section on page 289)

DIRECTIONS

1. Detangle your hair, comb it away from your face, and part it your favorite way.

2. Prepare the light blonde color according to the package instructions and put it in a plastic bowl.

3. Using the Tung brush, apply it to your hair in 1-inch wide ribbons (see the ribboning technique on page 47), starting 1½ inches down from your scalp through to the ends, all around your part and hairline. Be bold in your application!

4. Layer the unrolled cotton balls over the ribbons and lay plastic wrap on top.

5. Set the timer for 10 minutes and heat the ends with your hair dryer and diffuser, working in a circular motion and counting to 20 for each section.

6. When time is up, rinse, gently shampoo with no scalp manipulation, and lightly condition your hair. Towel blot your hair and detangle.

7. Apply the triple-barrier cream and put on gloves.

8. Take the developer bottle out of the auburn brown box, snip off its applicator tip, and remove the cap.

9. Add the auburn brown tint to the developer.

10. Shake it well to combine the materials, keeping your gloved finger over the open tip.

11. Apply the mixture to your hair starting from your scalp down through to the ends.

12. Comb through your hair to make sure the mixture is distributed evenly.

13. As soon as you are done with the application, saturate cotton balls with aerosol hair spray as needed and wipe away the triple-barrier cream in the direction of the hairline to avoid creating a harsh line. This ensures you will not stain your skin.

14. Set the timer for 20 minutes.

15. When time is up, rinse your hair, shampoo lightly with no scalp manipulation, and gently condition. Towel dry your hair and style as usual.

16. Use Colorist Collection Mahogany once a week to help maintain the depth of your color.

MATERIALS

1 box light blonde

Plastic bowl

Tung brush

Homemade water-jug paddle

8 unrolled cotton balls

Plastic wrap

Timer

Hair dryer and diffuser or sock diffuser

Shampoo and conditioner

Triple-barrier cream

1 box auburn brown (I like L'Oréal Paris Féria #56)

Measuring cups and spoons

8 cotton balls

Aerosol hair spray

L'Oréal Paris Colorist Collection Mahogany Shampoo*

Available at beauty supply stores and online (see the Resources section on page 289)

DIRECTIONS

1. Detangle your hair and comb it away from your face. Part your hair in your favorite place.

2. Prepare the light blonde color according to the package directions and put it in a plastic bowl.

3. Using your Tung brush and water-jug paddle, apply it to your hair in 1-inch ribbons all around your part and hairline, starting 1½ inches down from your scalp through the ends (see the ribboning technique on page 47). Be bold in your application!

4. As soon as you are done, lay the unrolled cotton balls on the ribbons. Layer plastic wrap on top.

5. Set the timer for 5 minutes. Apply high heat from your hair dryer and diffuser, concentrating on the ends and counting to 20 for each ribbon.

6. When time is up, rinse, gently shampoo with no scalp manipulation, and lightly condition your hair. Towel blot your hair well and detangle it.

7. Apply the triple-barrier cream and put on gloves.

8. Take the developer bottle out of the auburn brown box, snip off its applicator tip, and remove the cap.

9. Add the auburn brown tint to the developer.

10. Shake the bottle well to combine the materials, keeping your gloved finger over the open tip.

11. Apply the mixture to your hair starting from your scalp down through to the ends.

12. Comb through your hair to make sure the mixture is distributed evenly.

13. As soon as you are done with the application, saturate cotton balls with aerosol hair spray as needed and wipe away the triple-barrier cream in the direction of the hairline to avoid creating a harsh line. This ensures you will not stain your skin.

14. Set the timer for 20 minutes.

15. When time is up, rinse your hair, shampoo lightly with no scalp manipulation, and gently condition. Towel dry and style as usual.

16. Use Colorist Collection Mahogany once a week to help maintain the depth of your color.

Tab 5

MATERIALS

1 box light blonde

Plastic bowl

Tung brush

Homemade water-jug paddle

8 unrolled cotton balls

Plastic wrap

Timer

Hair dryer and diffuser or diffuser sock

Shampoo and conditioner

Triple-barrier cream

Distilled water

1 box auburn brown

8 cotton balls

Aerosol hair spray

L'Oréal Paris Colorist Collection Mahogany Shampoo*

Available at beauty supply stores and online (see the Resources section on page 289)

DIRECTIONS

1. Detangle your hair and comb it away from your face. Part your hair in your favorite place.

2. Prepare the light blonde color according to the package directions and put it in a plastic bowl.

3. Using your Tung brush and water-jug paddle, apply it to your hair in 1-inch ribbons all around your part and hairline, starting 1½ inches down from your scalp through the ends (see the ribboning technique on page 47). Be bold in your application!

4. As soon as you are done, lay the unrolled cotton balls on the ribbons. Layer plastic wrap on top.

5. Set the timer for 10 minutes. Apply high heat from your hair dryer and diffuser, concentrating on the ends, using a circular motion, and counting to 20 for each ribbon.

6. When time is up, rinse, shampoo, detangle your hair, and towel blot well.

7. Take the developer out of the auburn brown kit, snip off the application tip, and remove the cap.

8. Using measuring cups or spoons, remove and discard 1 ounce (½ developer container) of the developer in the auburn brown kit. Replace it with 1 ounce of distilled water.

9. Add the auburn brown tint to the developer.

10. Shake the bottle well to combine the materials, keeping your gloved finger over the open tip.

11. Apply the mixture to your hair starting from your scalp through to the ends.

12. Comb through your hair to make sure the mixture is distributed evenly.

13. As soon as you are done with the application, saturate cotton balls with aerosol hair spray as needed and wipe away the triple-barrier cream in the direction of the hairline to avoid creating a harsh line. This ensures you will not stain your skin.

14. Set timer for 20 minutes.

15. When time is up, apply the remainder starting from your scalp to 1½ inches down, and comb the mixture through your hair to make sure it is distributed evenly.

16. Set the timer for 10 minutes.

17. When time is up, rinse your hair, shampoo gently with no scalp manipulation, gently condition, and towel dry. Style as usual.

18. Use Colorist Collection in Mahogany once a week to help maintain the depth of your color.

Tabs 5½ and 6

MATERIALS

1 box light blonde

Plastic bowl

Tung brush

Homemade water-jug paddle

8 unrolled cotton balls

Plastic wrap

Timer

Shampoo and conditioner

Triple-barrier cream

1 box auburn brown (I like L'Oréal Paris Féria #56)

Measuring cups and spoons

Distilled water

8 cotton balls

Aerosol hair spray

L'Oréal Paris Colorist Collection Mahogany Shampoo*

Available at beauty supply stores and online (see the Resources section on page 289)

DIRECTIONS

1. Detangle your hair and comb it away from your face. Part your hair in your favorite place.

2. Prepare the light blonde color according to the package directions and put it in a plastic bowl.

3. Using your Tung brush and water-jug paddle, apply it to your hair in 1-inch ribbons all around your part and hairline, starting 1½ inches down from your scalp through the ends (see the ribboning technique on page 47). Be bold in your application!

4. As soon as you are done, lay the unrolled cotton balls on the ribbons. Layer plastic wrap on top.

5. Set the timer for 10 minutes.

6. When time is up, rinse, shampoo, detangle your hair, and towel blot well.

7. Take the developer out of the auburn brown kit, snip off the application tip, and remove the cap.

8. Using measuring cups or spoons, remove about ½ the developer (be sure to note the exact amount) and discard. Add that same amount of distilled water. Add the auburn brown tint to the developer.

9. Shake the bottle until the materials are blended, keeping your gloved finger over the open tip.

10. Apply it to your hair, starting from your scalp through to the ends. Comb through to ensure even application.

11. As soon as you are done with the application, saturate cotton balls with aerosol hair spray as needed and wipe away the triple-barrier cream in the direction of the hairline to avoid creating a harsh line. This ensures you will not stain your skin.

12. Set the timer for 20 minutes.

13. When time is up, rinse, shampoo, and lightly condition your hair. Towel blot and detangle. Style as usual.

14. Use Colorist Collection Mahogany once a week to maintain the depth of your color.

MATERIALS

Triple-barrier cream

1 box auburn brown (I like L'Oréal Paris Féria #56)

Measuring cups and spoons

Distilled water

8 cotton balls

Aerosol hair spray

Timer

Shampoo and conditioner

L'Oréal Paris Colorist Collection Mahogany Shampoo*

Available at beauty supply stores and online (see the Resources section on page 289)

DIRECTIONS

1. Detangle your hair and comb it away from your face.

2. Apply the triple-barrier cream and put on gloves.

3. Snip tip of developer applicator and remove the cap. Using measuring cups or spoons, remove about ½ the developer (be sure to note the exact amount). Add that same amount of distilled water. Add the auburn brown tint to the developer.

4. Shake the bottle until materials are blended, keeping your gloved finger over the open tip.

5. Apply the mixture to your hair from your scalp down to the ends.

6. As soon as you are done with the application, saturate cotton balls with aerosol hair spray as needed and wipe away the triple-barrier cream in the direction of the hairline to avoid creating a harsh line. This ensures you will not stain your skin.

7. Set the timer for 20 minutes.

8. When time is up, rinse your hair, shampoo well, condition, and towel dry. Style as usual.

9. Use Colorist Collection in Mahogany once a week to maintain the depth of your color.

Tabs 7 and 8

Follow your tab recipe for Milk Chocolate or Cream-Swirled Milk Chocolate and follow the Cherry Cola gloss recipe for Tabs 5½ and 6 (see page 122). Use L'Oréal Paris Colorist Collection Red Mahogany Shampoo once a week to help maintain the depth of your color.

Tab 9

Revert to your pre-gray natural tab recipe. However, if you really want more coverage, use Tab 11's Milk Chocolate recipe first and add the Auburn Brown gloss from the recipe for Tabs 5½ and 6 (see page 122).

Tab 10

If you are closer to 35 percent gray, revert to your pre-gray natural tab recipe. However, if you really want more coverage, use your pre-gray tab's Milk Chocolate recipe first and add the Auburn Brown gloss from the recipe for Tabs 5½ and 6 (see page 122). If you are closer to 65 percent gray, use the recipe for Tab 11.

Tab 11

MATERIALS

1 perm kit

Triple-barrier cream

1 box dark golden brown

1 box auburn

1 box L'Oréal Paris Féria Deep Copper

Gray Magic*

8 cotton balls

Aerosol hair spray

Timer

Shampoo and conditioner

1 box auburn brown

Distilled water

L'Oréal Paris Colorist Collection Mahogany Shampoo*

*Available at beauty supply stores and online (see the Resources section on page 289)

DIRECTIONS

1. Pre-soften hair: Apply the perm solution to your hair starting about 1¼ inches from the scalp down to the ends and let it work for 3 minutes. Rinse, shampoo with no scalp manipulation, lightly comb out your hair, and towel blot. You are ready to continue with your recipe.

2. Apply the triple-barrier cream and put on gloves.

3. Snip off the applicator tip of the developer and remove the cap.

4. Add the whole color container of dark golden brown to the developer.

5. Using a measuring spoon, add 1 teaspoon of auburn.

6. Add 1 teaspoon of deep copper.

7. Add 2 drops of Gray Magic.

8. Shake the mixture until the materials are combined, holding a gloved finger over the open tip.

9. Apply the mixture to your hair from the scalp through to the ends. Comb through so that the mixture is distributed evenly.

10. As soon as you are done with the application, saturate cotton balls with aerosol hair spray as needed and wipe away the triple-barrier cream in the direction of the hairline to avoid creating a harsh line. This ensures you will not stain your skin.

11. Set the timer for 25 minutes.

12. Shampoo gently with no scalp manipulation, condition lightly to detangle, and towel blot well.

13. Snip the tip off the developer bottle and remove the cap.

14. Using measuring cups and spoons, remove half of the developer and replace it with distilled water.

15. Put on gloves and add the auburn brown tint to the developer bottle.

16. Shake the bottle to combine the tints and developer, keeping your gloved finger over the open tip.

17. Apply the mixture to your hair from the scalp all the way through to the ends. Comb well to ensure even coverage.

18. Set the timer for 15 minutes.

19. When time is up, rinse your hair and shampoo well. Do a second shampoo with Colorist Collection Mahogany Shampoo, condition, towel dry, and style as usual.

20. Use the Mahogany Shampoo for every third shampoo to help extend your color.

6

BEAUTIFULLY BURNISHED BRUNETTES

These are recipes for light brown and soft, deep blonde shades. They are romantic, subtle, and surprisingly sensual.

Target colors are:

- **TOASTED CINNAMON**

- **TOASTED CINNAMON WITH BUTTERED LIGHTS**

- **DEEP CARAMEL**

- **DEEP CARAMEL WITH VANILLA RIBBONS**

- **TOASTED WHEAT WITH VANILLA THREADS**

NOTE: If you have very thick, long hair (shoulder length or longer), double all recipes.

NOTE: If you have very short hair, use the Frosted Secret Sauce technique on page 247 (in the Just for Men chapter—*not!*) to highlight your hair.

TOASTED CINNAMON

Tab 1

Triple-barrier cream

2 clean washcloths

1 box light blonde

1 box reddish brown (I prefer Clairol Nice'n Easy or Revlon)

8 cotton balls

Aerosol hair spray

Timer

Shampoo and conditioner

L'Oréal Paris Colorist Collection Walnut Shampoo*

*Available at beauty supply stores and online (see the Resources section on page 289)

READ ME! PREPARATION IS EVERYTHING

Before you begin any recipe, read through the ingredient list, get out your color kit, and prepare your work area. Clear off the countertop on which you will be working and spread out your materials. Detangle your hair without manipulating the scalp. Put on your work shirt and wrap a towel around your neck to catch stains. Apply the triple-barrier cream around your hairline (page 37), especially for brunette recipes. Cover the floor with a drop cloth or old newspapers. (Remember: no plastic underfoot!) Be sure to have your timer out and ready to set. *Wear kit-provided gloves when applying color!* You can recycle gloves by rinsing them inside and out, patting them dry, and sprinkling some baby powder or cornstarch on the inside for easy slip-on next time.

---

1. Detangle your hair and comb it away from your face.

2. Prepare the light blonde tint according to the package instructions.

3. Apply the mixture to your hair starting 1½ inches down from your scalp.

4. Set the timer for 10 minutes.

5. When time is up, rinse your hair, shampoo gently without scalp manipulation, and blot well with a towel.

6. Apply the triple-barrier cream.

7. Prepare the reddish brown tint according to the package instructions.

8. Apply the mixture to your hair from your scalp to the ends and comb through your hair to make sure the mixture is distributed evenly.

9. As soon as you are done with the application, saturate cotton balls with aerosol hair spray as needed and wipe away the triple-barrier cream in the direction of the hairline to avoid creating a harsh line. This ensures you will not stain your skin.

10. Set the timer for 30 minutes.

11. When time is up, rinse your hair and shampoo gently.

12. Do a second shampoo with Colorist Collection Walnut Shampoo, lightly condition your hair, and style as usual.

Tabs 2, 3, and 4

MATERIALS

Measuring cups and spoons

1 box light blonde tint

Timer

Shampoo and conditioner

Triple-barrier cream

Measuring cups and spoons

1 box medium golden brown

1 box reddish brown

8 cotton balls

Aerosol hair spray

L'Oréal Paris Colorist Collection Walnut Shampoo*

*Available at beauty supply stores and online (see the Resources section on page 289)

DIRECTIONS

1. Detangle your hair and comb it away from your face.

2. Mix the light blonde kit per the package instructions.

3. Apply the mixture 1½ inches from the scalp through to the ends. Comb through carefully to ensure even distribution.

4. Set the timer for 5 minutes.

5. When time is up, rinse, shampoo well, detangle, and towel blot.

6. Apply the triple-barrier cream and put on gloves.

7. Take the developer out of the box of medium golden brown tint, snip off applicator tip, and remove cap.

8. Using measuring cups or spoons, add ¼ ounce (⅛ color container) of the medium golden brown tint to the developer.

9. Add 1¾ ounces (three ¼ + ⅛ color container) of the reddish brown tint to the developer.

10. Gently shake the bottle to mix the materials, holding your gloved finger over the open tip.

11. Apply approximately ¾ the mixture to your hair starting 1½ inches down from your scalp through to the ends. Comb through your hair to make sure the mixture is distributed evenly.

12. Set the timer for 10 minutes.

13. When time is up, apply the remainder of the mixture starting from your scalp to 1½ inches down and comb through your hair to make sure mixture is distributed evenly.

14. As soon as you are done with the application, saturate cotton balls with aerosol hair spray as needed and wipe away the triple-barrier cream in the direction of the hairline to avoid creating a harsh line. This ensures you will not stain your skin.

15. Leave the mixture in for 10 minutes more.

16. When time is up, rinse your hair and shampoo gently. Shampoo again using Colorist Collection Walnut Shampoo, condition, and style as usual.

Tab 5

MATERIALS

Triple-barrier cream

Measuring cups and spoons

1 box light golden brown

1 box reddish brown

1 box deep copper (such as L'Oréal Paris Féria, Clairol Nice'n Easy, or Revlon)

Timer

8 cotton balls

Aerosol hair spray

Shampoo and conditioner

L'Oréal Paris Colorist Collection Walnut Shampoo*

*Available at beauty supply stores and online (see the Resources section on page 289)

1. Detangle your hair and comb it away from your face.

2. Apply the triple-barrier cream and put on gloves.

3. Take the developer out of the box of golden brown, snip off applicator tip, and remove the cap.

4. Using measuring cups or spoons, take 1 ounce (½ color container) of light golden brown tint and add it to the developer.

5. Add 1 ounce (½ color container) of reddish brown tint to the developer.

6. Add 1 teaspoon deep copper.

7. Gently shake the bottle to combine the materials, holding your gloved finger over the open tip.

8. Apply approximately ¾ the mixture to your hair starting 1½ inches down from your scalp through to the ends.

9. Leave the mixture on for 10 minutes and apply the remainder of the mixture starting from your scalp to 1½ inches down. Comb through your hair to make sure the mixture is distributed evenly.

10. As soon as you are done with the application, saturate cotton balls with aerosol hair spray as needed and wipe away the triple-barrier cream in the direction of the hairline to avoid creating a harsh line. This ensures you will not stain your skin.

11. Leave the mixture in for 20 minutes.

12. When time is up, rinse your hair and shampoo gently. Shampoo again using Colorist Collection Walnut Shampoo, lightly condition your hair, and style as usual.

Tabs 5½ and 6

Triple-barrier cream

1 box light golden brown

1 box light reddish brown (such as L'Oréal Paris Féria, Clairol Nice'n Easy, or Revlon)

1 box copper

Measuring cups and spoons

Timer

8 cotton balls

Aerosol hair spray

Shampoo and conditioner

L'Oréal Paris Colorist Collection Walnut Shampoo*

*Available at beauty supply stores and online (see the Resources section on page 289)

DIRECTIONS

1. Detangle your hair and comb it away from your face.

2. Apply the triple-barrier cream and put on gloves.

3. Take the developer out of the box of light golden brown, snip off applicator tip, and remove the cap.

4. Using measuring cups or spoons, add 1¾ ounces (three ¼ + ⅛ color container) of the light golden brown tint to the developer.

5. Add ¼ ounce (⅛ color container) of the light reddish brown tint to the developer.

6. Add 2 teaspoons of the deep copper.

7. Gently shake the bottle to combine the materials, holding your gloved finger over the open tip.

8. Apply approximately ¾ the mixture to your hair starting 1½ inches from your scalp to the ends.

9. Set the timer for 5 minutes.

10. When time is up, apply the remainder of the mixture to your hair starting from your scalp to 1½ inches down.

11. Set the timer for 25 minutes.

12. As soon as you are done with the application, saturate cotton balls with aerosol hair spray as needed and wipe away the triple-barrier cream in the direction of the hairline to avoid creating a harsh line. This ensures you will not stain your skin.

13. When time is up, rinse your hair and shampoo well. Do a second shampoo using Colorist Collection Walnut Shampoo, lightly condition your hair, and style as usual.

Tab 6½

MATERIALS

Triple-barrier cream

Measuring cups and spoons

1 box light reddish brown

1 box light golden brown (such as L'Oréal Paris Féria, Clairol Nice'n Easy, or Revlon)

8 cotton balls

Aerosol hair spray

Timer

Shampoo and conditioner

L'Oréal Paris Colorist Collection Walnut Shampoo*

*Available at beauty supply stores and online (see the Resources section on page 289)

1. Detangle your hair and comb it away from your face.

2. Apply the triple-barrier cream and put on gloves.

3. Remove the developer from the box of light reddish brown, snip off the applicator tip, and remove the cap.

4. Using measuring cups or spoons, add ½ ounce (¼ color container) of the light reddish brown tint to developer.

5. Add 1½ ounces (three ¼ color container) of the light golden brown tint to the developer.

6. Shake the bottle gently to combine the materials, holding your gloved finger over the open tip.

7. Apply the mixture to your hair from your scalp through the ends.

8. As soon as you are done with the application, saturate cotton balls with aerosol hair spray as needed and wipe away the triple-barrier cream in the direction of the hairline to avoid creating a harsh line. This ensures you will not stain your skin.

9. Set the timer for 30 minutes.

10. When time is up, rinse your hair and shampoo gently. Shampoo again using Colorist Collection Walnut Shampoo, lightly condition your hair, and style as usual.

Tabs 7 and 8

MATERIALS

Triple-barrier cream

Measuring cups and spoons

1 box medium golden brown (if you can't find, combine equal amounts dark golden brown and light golden brown)

1 box dark golden brown

1 box deep copper (such as L'Oréal Paris Féria, Clairol Nice'n Easy, or Revlon)

8 cotton balls

Aerosol hair spray

Timer

Shampoo and conditioner

L'Oréal Paris Colorist Collection Walnut Shampoo*

*Available at beauty supply stores and online (see the Resources section on page 289)

DIRECTIONS

1. Detangle your hair and comb it away from your face.

2. Apply the triple-barrier cream and put on gloves.

3. Remove the developer from the box of medium golden brown, snip off the applicator tip, and remove the cap.

4. Using measuring cups or spoons, add 1¾ ounces (three ¼ + ⅛ color container) of the medium golden brown to the developer.

5. Add ⅛ ounce (¹⁄₁₆ color container) of the dark golden brown to the developer.

6. Add ⅛ ounce (¹⁄₁₆ color container) of the deep copper.

TRIPLE-BARRIER CREAM—A REMINDER

First, with your fingertips apply a layer of lotion, such as Nivea or Keri brand, around your face at your hairline. Follow that with a layer of petroleum jelly. Then along the outside border of the lotion/petroleum jelly strip, apply a layer of Carmex lip balm.

7. Shake the bottle to combine the materials, holding your gloved finger over the open tip.

8. Apply the mixture to your hair starting from your scalp through the ends.

9. As soon as you are done with the application, saturate cotton balls with aerosol hair spray as needed and wipe away the triple-barrier cream in the direction of the hairline to avoid creating a harsh line. This ensures you will not stain your skin.

10. Set the timer for 25 minutes.

11. When time is up, rinse your hair and shampoo gently. Shampoo again using Colorist Collection Walnut Shampoo, lightly condition your hair, and style as usual.

Tab 9

Refer to the recipe for your pre-gray natural tab.

> **NOTE:** If you use Nioxin, it may offset some tints and impact color. Before you tint or highlight, do not use Nioxin for three days prior to any hair-color service. (People with thinning hair use Nioxin.)

Tab 10

If your hair is closer to 35 percent gray, use your pre-gray tab. If your hair is closer to 65 percent gray, use the recipe for Tab 11.

Tab 11

MATERIALS

1 perm kit

Triple-barrier cream

Measuring cups and spoons

1 box dark golden brown

1 box deep/intense copper (such as L'Oréal Paris Féria, Clairol Nice'n Easy, or Revlon)

8 cotton balls

Aerosol hair spray

Timer

Shampoo and conditioner

L'Oréal Paris Colorist Collection Walnut Shampoo*

*Available at beauty supply stores and online (see the Resources section on page 289)

DIRECTIONS

1. Pre-soften hair: Apply the permanent solution from the a drugstore-perm kit to your hair starting about 1¼ inches from your scalp down to the ends. Let it sit 1 minute, rinse, shampoo, and comb out your hair. You are ready to continue with the recipe.

2. Apply the triple-barrier cream and put on gloves.

3. Remove the developer from the box of dark golden brown, snip off the applicator tip, and remove the cap.

4. Using measuring cups or spoons, add 1¾ + ⅛ ounces (three ¼ + ⅛ + ¹⁄₁₆ color container) of the dark golden brown tint to the developer.

5. Add ⅛ ounce (¹⁄₁₆ color container) of copper tint.

6. Shake the bottle to combine the materials, holding your gloved finger over the open tip.

7. Apply the mixture to your hair, starting from your scalp through to the ends.

8. As soon as you are done with the application, saturate cotton balls with aerosol hair spray as needed and wipe away the triple-barrier cream in the direction of the hairline to avoid creating a harsh line. This ensures you will not stain your skin.

9. Set the timer for 30 minutes.

10. When time is up, rinse your hair and shampoo well. Shampoo again using Colorist Collection Walnut Shampoo, lightly condition your hair, and style as usual.

TOASTED CINNAMON
WITH BUTTERED LIGHTS

After your Toasted Cinnamon is completely dry, why not spread a little butter on it? It's an easy two-step process. Those of you with naturally light reddish-brown hair can use just the highlight formula to add some buttered lights.

Tabs 1, 2, 3, and 4

MATERIALS

1 box Clairol Frost'n Tip highlighting kit

Measuring cups and spoons

40-volume peroxide*

Rattail comb

5–7 yoyettes

5–7 pieces 3"x6" heavy-duty foil

Tung brush

Oral-B soft toothbrush with an elongated end

1 box light golden brown

Distilled water

Timer

Shampoo

*Available at beauty supply stores and online (see the Resources section on page 289)

DIRECTIONS

1. Complete the Toasted Cinnamon recipe for Tab 1, 2, 3, or 4 and dry hair completely.

2. Mix the Clairol Frost'n Tip kit according to the package directions, except remove 1 tablespoon of the developer and replace it with 1 tablespoon of 40-volume peroxide.

3. Part your hair in your favorite way.

4. Using your rattail comb, pull ¼-inch sections of hair (about 12 to 17 strands) along each side of your part, staggering the sections on each side. Clip them up with the yoyettes. (See the foil-out technique on page 41.)

5. Unclip one section at a time. For each, take a foil sheet and place it under the section. The hair should sit flat on the foil. Hold the section down with one hand.

6. Dip the Tung brush into the blonde mixture and spread the mixture over the hair.

7. Fold the foil around your hair and continue to the next section until all the sections are done on both sides of the part. Do not fold the foil too tightly. This is a holding and heating procedure.

8. Next, take the Oral-B toothbrush, dip it into the blonde mixture, and brush it on yarn-size strands around your face, starting around 1½ inches down from your scalp and keeping the application bottom heavy. It does not have to be neat—we are adding texture, so perfection is not required. (See the balayage technique on page 49.)

9. Set the timer for 20 minutes.

10. When time is up, rinse your hair and shampoo gently. Do not scrub your scalp, but do pay attention to your hairline and make sure you get all the product out.

11. Towel blot and detangle your hair.

12. Take the box of light golden brown, snip off the tip of the developer bottle, and remove the cap.

13. Using measuring cups or spoons, remove 1 ounce of developer (½ developer container). Replace it with 1 ounce of distilled water. Add the tint to the developer and shake the bottle until the materials are combined, placing your gloved finger over the open tip.

14. Apply the mixture to your hair from your scalp down to the ends and comb through your hair to make sure the mixture is distributed evenly.

15. Set the timer for 2 minutes.

16. When time is up, rinse your hair, shampoo gently with no scalp manipulation, and condition lightly. Towel dry and style as usual.

Tabs 5, 5½, 6, 6½, 7, 7½, 8, 9, 10, and 11

Complete your Toasted Cinnamon recipe and add the following highlight and warm gloss.

MATERIALS

1 box Clairol Frost'n Tip highlighting kit

Rattail comb

5–7 yoyettes

5–7 pieces 3"x6" heavy-duty foil

Tung brush

Oral-B soft toothbrush with an elongated end

1 box light golden brown

Distilled water

Timer

Shampoo and conditioner

DIRECTIONS

1. Complete the Toasted Cinnamon recipe for your tab and dry hair completely.

2. Mix the Clairol Frost'n Tip kit according to the package directions.

3. Part your hair in your favorite way.

4. Using your rattail comb, pull ¼-inch sections of hair (about 12 to 17 strands) along each side of your part, staggering sections on each side. Clip them up with the yoyettes. (See the foil-out technique on page 41.)

5. Unclip one section at a time. For each, take a foil sheet and place it under the section. The hair should sit flat on the foil. Hold the section down with one hand.

6. Dip the Tung brush into the blonde mixture and spread the mixture over the hair on the foil.

7. Fold the foil around the hair and continue to the next section until all the sections are done on both sides of the part. Do not fold the foil too tightly. This is simply a holding and heating technique.

8. Next, take the Oral-B toothbrush, dip it into the blonde mixture, and brush it on yarn-size strands around your face starting 1½ inches down from your scalp and keeping the application bottom heavy. It does not have to be neat—we are adding texture, so perfection is not required. (See the balayage technique on page 49.)

9. Set the timer for 10 minutes.

10. When time is up, rinse your hair and shampoo gently with no scalp manipulation.

11. Towel blot and detangle your hair.

12. Take the box of light golden brown, snip off the tip of the developer bottle, and remove the cap.

13. Using measuring cups or spoons, remove 1 ounce of developer (½ developer container) and replace it with 1 ounce of distilled water. Add the tint to the developer.

14. Shake the bottle gently until the materials are combined, placing your gloved finger over the open tip.

15. Apply the mixture to your hair quickly, starting from the scalp down to the ends, and comb through your hair to make sure it is distributed evenly.

16. Set the timer for 2 minutes.

17. When time is up, rinse your hair, shampoo gently with no scalp manipulation, and condition lightly. Towel dry and style as usual.

DEEP CARAMEL

This luscious recipe gives your hair a color as soft and sumptuous as the name implies.

Tabs 1, 2, 3, and 4

Deep Caramel is not applicable for Tabs 1, 2, 3, and 4. Try a Milk Chocolate highlight instead! Use your Milk Chocolate recipe with foil-out technique (see page 41)—just put your recipe tint in the foil! Rich, gorgeous, and easy on the eyes.

> **NOTE:** If your hair begins to look faded, you can use a combination of half L'Oréal Paris Colorist Collection Walnut Shampoo and half your regular shampoo every third cleansing.

Tabs 5 and 5½

MATERIALS

Triple-barrier cream

1 box light blonde (I like L'Oréal Paris Féria #100)

Plastic wrap

Hair dryer and diffuser or diffuser sock

8 cotton balls

Aerosol hair spray

Timer

Shampoo and conditioner

L'Oréal Paris Colorist Collection Ginger Root Shampoo*

*Available at beauty supply stores and online (see the Resources section on page 289)

DIRECTIONS

1. Detangle your hair and comb it away from your face.

2. Apply the triple-barrier cream and put on gloves.

3. Take the developer out of the light blonde kit, snip off the applicator tip, and remove the cap.

4. Mix the light blonde tint per the manufacturer's instructions.

5. Shake the bottle to mix, keeping your gloved finger on top of the open tip.

6. Apply approximately ¾ the mixture to your hair starting 1½ inches down from your scalp through to the ends. Comb through your hair to make sure the mixture is distributed evenly.

7. Place the plastic wrap on top of the color-saturated hair and heat for 3 minutes with the hair dryer and diffuser or diffuser sock. Remove the plastic wrap.

8. Apply the remainder of the mixture from the scalp to 1½ inches down, and comb through your hair again to make sure the mixture is distributed evenly.

9. As soon as you are done with the application, saturate cotton balls with aerosol hair spray as needed and wipe away the triple-barrier cream in the direction of the hairline to avoid creating a harsh line. This ensures you will not stain your skin.

10. Let the mixture work for 10 minutes.

11. When time is up, rinse your hair and shampoo gently. Use Colorist Collection Ginger Root Shampoo for a second shampoo immediately after the first. Condition, towel dry, and style as usual.

Tab 6 and 6½

MATERIALS

Triple-barrier cream

1 box light blonde (I like L'Oréal Paris Féria #100)

Timer

8 cotton balls

Aerosol hair spray

Shampoo and conditioner

L'Oréal Paris Colorist Collection Ginger Root Shampoo*

Available at beauty supply stores and online (see the Resources section on page 289)

DIRECTIONS

1. Detangle your hair and comb it away from your face.

2. Apply the triple-barrier cream and put on gloves.

3. Prepare the color according to the package directions.

4. Apply approximately ¾ the mixture to your hair starting 1½ inches down from your scalp through to the ends.

5. Leave the mixture in for 5 minutes.

6. Apply the remainder of the mixture to your hair starting from your scalp to 1½ inches down and comb through your hair to make sure the mixture is distributed evenly.

7. As soon as you are done with the application, saturate cotton balls with aerosol hair spray as needed and wipe away the triple-barrier cream in the direction of your hairline to avoid creating a harsh line. This ensures you will not stain your skin.

8. Leave the mixture in for 3 minutes.

9. When time is up, rinse your hair, shampoo well, and shampoo a second time with Colorist Collection Ginger Root Shampoo. Condition, towel blot, and style as usual.

Tabs 7 and 7½

MATERIALS

1 box light golden blonde

Shampoo and conditioner

L'Oréal Paris Colorist Collection Ginger Root Shampoo*

Available at beauty supply stores and online (see the Resources section on page 289)

1. Detangle your hair and comb it away from your face.

2. Prepare the color according to the package instructions.

3. Apply approximately ¾ the mixture to your hair starting 1½ inches from your scalp down through to the ends. Comb through your hair to make sure the mixture is distributed evenly. Work as quickly as possible.

4. Apply the remainder of the mixture to your hair starting from your scalp to 1½ inches down. Comb through your hair to make sure it is distributed evenly. Work as quickly as possible.

5. Immediately rinse your hair and shampoo well. Use Colorist Collection Ginger Root Shampoo for a second shampoo. Condition, towel dry, and style as usual.

Tab 8

I do not recommend Deep Caramel as an all-over color for Tab 8s. Instead, try creating some panels underneath your hair using the Peekaboo Punk technique (see page 236) with the recipe for Tabs 10 and 11 (see page 149). Follow with two shampoos using Colorist Collection Ginger Root Shampoo (see Resources, page 289).

Tab 9

Revert to your pre-gray tab.

THE TOO-BLONDE BLUES

I've found the Deep Caramel recipe for Tabs 10 and 11 great for people who are naturally blonde, but also for those who have overbleached hair. It will help move you back to your natural color while camouflaging the overdone blondeness. If you are overly bottle blonde and have been doing a one-process color (not an all-over bleach) for a long time, you must do a pre-color (remember, take out 1 ounce of developer and replace it with 1 ounce of distilled water) of deep copper and let it work for 30 minutes. Also, always do a strand test to make sure your color will take. And remember, for major color correction, see a professional.

Tabs 10 and 11

MATERIALS

1 perm kit

Triple-barrier cream

1 box light golden brown

Measuring cups and spoons

1 box light reddish blonde

8 cotton balls

Aerosol hair spray

Timer

Shampoo and conditioner

L'Oréal Paris Colorist Collection Ginger Root Shampoo*

Available at beauty supply stores and online (see the Resources section on page 289)

DIRECTIONS

1. Pre-soften hair: Apply the permanent solution to your hair starting about 1¼ inches from the scalp down to the ends and let it work for 1 minute. Rinse, shampoo, and

condition. Comb out your hair and towel blot well. You are ready to continue with your recipe.

2. Apply the triple-barrier cream and put on gloves.

3. Remove the developer from the light golden brown kit, snip off the applicator tip, and remove the cap.

4. Using measuring cups or spoons, add 1½ ounces (three ¼ color container) light golden brown tint to the developer.

5. Add ½ ounce (¼ color container) of light reddish blonde tint to the developer.

6. Shake the bottle to mix the materials, holding your gloved finger over the open tip.

7. Apply the mixture to your hair starting from your scalp down through to the ends. Comb through to make sure the color is evenly distributed.

8. As soon as you are done with the application, saturate cotton balls with aerosol hair spray as needed and wipe away the triple-barrier cream in the direction of the hairline to avoid creating a harsh line. This ensures you will not stain your skin.

9. Set the timer for 35 minutes.

10. When time is up, rinse your hair and shampoo well. Shampoo a second time with Colorist Collection Ginger Root Shampoo, condition, towel dry, and style as usual.

DEEP CARAMEL WITH VANILLA RIBBONS

All Tabs

Add some extra flavor to your Deep Caramel (and if your hair is already this tawny tone, feel free to simply proceed with the highlights). These pretty ribbons of lightness add dimension to the overall color. It's so similar to adding a little vanilla to a cookie recipe. It makes all the difference to the final result.

This is a variation on the standard highlight recipe on page 44; refer back to that recipe for more details on materials and techniques.

MATERIALS

1 box Clairol Frost'n Tip Highlighting kit

Measuring cups and spoons

40-volume peroxide*

Rattail comb

5–7 yoyettes

5–7 pieces 3"x6" heavy-duty foil

Tung brush

Oral-B soft toothbrush with an elongated end

Timer

1 lemon (to match color)

Shampoo and conditioner

Clairol Shimmer Lights Shampoo*

*Available at beauty supply stores and online (see the Resources section on page 289)

DIRECTIONS

1. Complete your Deep Caramel color and make sure your hair is totally dry.

2. Mix the Clairol Frost'n Tip kit according to the package directions, except remove 1 tablespoon of the developer and replace it with 1 tablespoon of 40-volume peroxide.

3. Part your hair in your favorite way.

4. Using your rattail comb, pull ¼-inch sections of hair (about 12 to 17 strands) along each side of your part, staggering the sections on each side. Clip them up with the yoyettes. (See the foil-out technique on page 41.)

5. Unclip one section at a time. For each, take a foil sheet and place it under the section. The hair should sit flat on the foil. Hold the section down with one hand.

6. Dip the Tung brush into the blonde mixture and spread the mixture over the hair.

7. Fold the foil around your hair and continue to the next section until all the sections are done on both sides of the part. Do not fold the foil too tightly. This is a holding and heating procedure.

8. Next, take the Oral-B toothbrush, dip it into the blonde mixture, and brush it on yarn-size strands around your face, starting around 1½ inches down from your scalp and keeping the application bottom heavy. It does not have to be neat—we are adding texture, so perfection is not required. (See the balayage technique on page 49.)

9. Set the timer for 5 minutes.

10. When time is up, check one of the highlights to see the color. You're looking to match the color of the inside of a lemon rind (the white part). If you haven't achieved that shade, check again at 6 minutes and at 7 minutes—and don't go any longer.

11. Rinse your hair and shampoo it well. Do a second shampoo with Clairol Shimmer Lights Shampoo, which finishes the pigment and shines the color. It's kind of like using a tinted moisturizer on your face! Condition and style as usual.

TOASTED WHEAT
WITH VANILLA THREADS

This gorgeous shade draws inspiration from the Great Plains and prairies of America's heartland. When the sun shines on the fields of grain, the wheat gets "toasted" and takes on a deeper hue.

Tabs 1, 2, 3, 4, and 5

Toasted Wheat is not applicable for Tabs 1, 2, 3, 4, and 5. If you really crave this shade, you must see a professional!

Tabs 5½ and 6

MATERIALS

Triple-barrier cream

1 box dark ash blonde (I like Clairol Natural Instincts #10)

8 cotton balls

Aerosol hair spray

Timer

Shampoo and conditioner

DIRECTIONS

1. Detangle your hair and comb it away from your face.

2. Apply the triple-barrier cream and put on gloves.

3. Remove the developer from the kit, snip off the applicator tip, and remove the cap.

4. Add the dark ash blonde tint to the developer.

5. Shake the bottle to mix, holding your gloved finger over the open tip.

6. Apply the mixture to your hair, starting from the scalp through to the ends. Comb through your hair to make sure the mixture is distributed evenly.

7. As soon as you are done with the application, saturate cotton balls with aerosol hair spray as needed and wipe away the triple-barrier cream in the direction of the hairline to avoid creating a harsh line. This ensures you will not stain your skin.

8. Set the timer for 30 minutes.

9. When time is up, rinse your hair and shampoo gently with no scalp manipulation. Condition and towel dry.

10. Head to the Vanilla Threads recipe on page 160!

Tab 6½

Toasted Wheat is a bit dimming to Tab 6½. If you want to try it, however, try this recipe.

MATERIALS

Triple-barrier cream

1 box dark ash blonde

8 cotton balls

Aerosol hair spray

Timer

Shampoo and conditioner

DIRECTIONS

1. Detangle your hair and comb it away from your face.

2. Apply the triple-barrier cream and put on gloves.

3. Prepare the dark ash blonde tint according to package instructions.

4. Apply approximately ¾ the mixture to your hair from the scalp through to the ends and leave it in for 5 minutes.

5. Apply the remaining mixture starting from your scalp to 1½ inches down and comb through your hair.

6. As soon as you are done with the application, saturate cotton balls with aerosol hair spray as needed and wipe away the triple-barrier cream in the direction of the hairline to avoid creating a harsh line. This ensures you will not stain your skin.

7. Let the color work for 20 minutes.

8. Rinse, shampoo gently, and condition your hair.

10. Head to the Vanilla Threads recipe on page 160!

Tab 7, 7½, and 8

I suggest going straight to the recipe for Vanilla Threads, but you can deepen your color a bit with this recipe.

MATERIALS

Triple-barrier cream

1 box light golden brown

8 cotton balls

Aerosol hair spray

Timer

Shampoo and conditioner

1. Detangle your hair and comb it away from your face.

2. Remove the developer from the light golden brown box, snip off applicator tip, and remove the cap.

3. Add the light golden brown tint to the developer.

4. Shake the bottle gently to combine the materials, keeping your gloved finger over the open tip.

5. Apply the mixture to your hair starting from your scalp through to the ends.

6. As soon as you are done with the application, saturate cotton balls with aerosol hair spray as needed and wipe away the triple-barrier cream in the direction of your hairline to avoid creating a harsh line. This ensures you will not stain your skin.

7. Set the timer for 25 minutes.

8. When time is up, rinse your hair and shampoo gently with no scalp manipulation, condition, and towel dry. Style as usual.

10. Head to the Vanilla Threads recipe on page 160!

Tab 9

Revert to the recipe for your natural pre-gray tab color.

Tab 10

The beauty of gray is that you can go lighter than your natural color when coloring. For example, if you are more than 50 percent gray and your natural color is Tab 2 through 5, enjoy a lighter color. If your hair is closer to 35 percent gray, revert to your natural tab.

Triple-barrier cream

1 box dark blonde

Measuring cups and spoons

1 box medium golden brown

8 cotton balls

Aerosol hair spray

Timer

Shampoo and conditioner

DIRECTIONS

1. Detangle your hair and comb it away from your face.

2. Apply the triple-barrier cream and put on gloves.

3. Take the developer out of the dark blonde kit, snip off the applicator tip, and remove the cap.

4. Using measuring cups or spoons, add 1 ounce (½ color container) of dark blonde tint to the developer.

5. Add 1 ounce (½ color container) medium golden brown tint to the developer.

6. Shake the bottle to mix the materials, holding your gloved finger over open tip.

7. Apply the mixture to your hair starting from your scalp through to the ends. Comb through your hair to ensure the mixture is distributed evenly.

8. As soon as you are done with the application, saturate cotton balls with aerosol hair spray as needed and wipe away the triple-barrier cream in the direction of the hairline to avoid creating a harsh line. This ensures you that will not stain your skin.

9. Set the timer for 30 minutes.

10. When time is up, rinse your hair, shampoo well, condition, and towel dry.

11. Head to the Vanilla Threads recipe on page 160!

Tab 11

MATERIALS

1 perm kit

Triple-barrier cream

1 box medium golden brown

Measuring cups and spoons

1 box dark blonde

1 box deep copper

8 cotton balls

Aerosol hair spray

Timer

Shampoo and conditioner

DIRECTIONS

1. Pre-soften hair: Apply the permanent solution to your hair starting about 1¼ inches from the scalp down to the ends and let it work for 5 minutes. Rinse, shampoo, and comb out your hair. You are ready to continue with your recipe.

2. Apply the triple-barrier cream and put on gloves.

3. Take the developer out of the dark blonde kit, snip off applicator tip, and remove the cap.

4. Using measuring cups or spoons, add 1½ ounces (three ¼ color container) of the medium golden brown tint to the developer.

5. Add ½ ounce (¼ color container) of the dark blonde tint to the developer.

6. Add 1 teaspoon deep copper.

7. Shake the bottle to mix the materials, holding your gloved finger over open tip.

8. Apply mixture to your hair from your scalp through to the ends. Comb through to ensure the mixture is distributed evenly.

9. As soon as you are done with the application, saturate cotton balls with aerosol hair spray as needed and wipe away the triple-barrier cream in the direction of the hairline to avoid creating a harsh line. This ensures you will not stain your skin.

10. Set the timer for 25 minutes.

11. When time is up, rinse your hair, shampoo well, condition, and towel dry.

12. Head to the Vanilla Threads recipe on page 160.

VANILLA THREADS

All Tabs

This is a variation on the standard highlight recipe on page 44; refer to that recipe for more details on materials and techniques.

MATERIALS

1 box Clairol Frost'n Tip Dramatic Highlighting kit

Measuring cups and spoons

40-volume peroxide*

Oral-B soft toothbrush with an elongated end

Timer

1 banana (to match color)

Shampoo

Clairol Shimmer Lights Shampoo*

Available at beauty supply stores and online (see the Resources section on page 289)

DIRECTIONS

1. Complete your Toasted Wheat color and make sure your hair is totally dry.

2. Mix the Clairol Frost'n Tip kit according to the package directions, except remove 2 tablespoons of the developer and replace it with 2 tablespoons of 40-volume peroxide.

3. Part your hair in your favorite way.

4. Use the Oral-B toothbrush to place the mixture on strands around your hairline and on the hair around your part, starting ½ inch from the scalp all the way through the ends and keeping the application bottom heavy. (See the threading technique on page 46).

5. Set the timer for 25 minutes.

6. When time is up, check one of the highlights to see the color. You're looking to match the color of the inside of a banana peel.

7. When you've reached your desired color, rinse your hair, shampoo, and condition. You may want to use Clairol Shimmer Lights Shampoo once a week to maintain your highlights.

7.

RED-HOT REDHEADS

These recipes are for the ravishing reds—from wine colors to snappy gingers. These colors are very easily achieved; you can't go wrong with red. But you also have to be ready for red—it's a traffic-stopping color, literally! Can you handle the attention? Do you dare to be dramatic? If so, you're ready to ascend the crimson staircase!

Target colors are:

- **SMOKY PLUM WINE**

- **MERLOT**

- **SPICED PERSIMMON**

- **GINGERED TOFFEE**

- **APRICOT**

NOTE: If you have very thick, long hair (shoulder length or longer), double all recipes.

SMOKY PLUM WINE

This succulent color is quite intoxicating! It's deep yet subtle and just a *tiny* bit sweet. If you have very short hair, try this recipe with the Frosted Secret Sauce technique in the men's chapter, page 247.

Tabs 1, 2, and 3

MATERIALS

1 box Clairol Frost'n Tip highlighting kit

40-volume cream peroxide

Olive oil

9 yoyettes

Tung brush

Oral-B toothbrush, soft, with an elongated end

Heavy-duty foil

Triple-barrier cream

READ ME! PREPARATION IS EVERYTHING

Before you begin any recipe, read through the ingredient list, get out your color kit, and prepare your work area. Clear off the countertop on which you will be working and spread out the materials. Detangle hair without manipulating the scalp. Put on your work shirt, and wrap a towel around your neck and secure with a butterfly clip to catch stains. Apply the triple-barrier cream around your hairline (page 37), especially for brunette recipes. Cover the floor with a drop cloth or old newspapers. (Remember, no plastic underfoot!). Be sure to have your timer out and ready to set. *Wear kit-provided gloves when applying color!* You can recycle gloves by rinsing them inside and out, patting them dry and sprinkling some baby powder or cornstarch on the inside for easy slip-on next time.

L'Oréal Paris Chilled Plum Color Pulse Mousse

8 cotton balls

Aerosol hair spray

Timer

Color-friendly shampoo and conditioner (I like Garnier or L'Oréal Paris Color Vive)

DIRECTIONS

1. Prepare the highlighting mixture: Use the cap of the 40-volume peroxide as a measure and remove 2 capfuls from the peroxide bottle in the Frost'n Tip kit. Discard the liquid.

2. Wash the cap and use it to measure 2 capfuls of 40-volume peroxide. Add them to the Frost'n Tip peroxide bottle. Add 1½ teaspoons olive oil. Rinse the cap well and dry it thoroughly before replacing it on the bottle.

3. Part your dry hair the way you like to wear it best. First mark two 1-inch ribbons, 1 inch back from both sides of your face, and clip them with yoyettes.

4. Next, use yoyette clips to clasp two ¼-inch ribbons starting behind the 1-inch ribbon on the left side of your hair if you parted your hair on the left, or the right side of your hair if you parted your hair on the right.

5. Use yoyette clips to clasp three ¼-inch ribbons on the opposite side of your part, but not directly across from the other ribbons. The ribbons should look staggered, like a checkerboard. (See the foil-out technique on page 41.)

6. Hold the first 1-inch section in your hand, remove the clip, and place a piece of foil under it. Use the Tung brush to apply the highlighting mixture, starting about 1 inch from the scalp all the way down to the ends. Fold the foil around the hair to hold it in place. Do the same on the ribbon on the opposite side.

7. Now move on to the ¼-inch ribbons. Working on one side and then the other, hold each ribbon section in your hand, remove the clip, and place a piece of foil under it.

Use the Tung brush to apply the highlighting mixture, starting at your scalp all the way to the ends. Be sure to go bottom heavy with the highlights on these ribbons. As you finish each ribbon, fold the foil to hold the hair in place.

8. When you are through applying highlights, take the Oral-B toothbrush and dip it into the highlighting mixture and brush fine strands on either side of your face to frame it. If you have really long hair, take the Tung brush, dip it into the highlighting mixture, and brush the front ends of the hair with it.

9. Set the timer for 2 minutes. When the time is up, unfold one highlight and check the color. You want it to match a piece of milk chocolate. If you have that color, unwrap all the highlights and rinse your hair well, then towel blot and detangle. If you have not reached the right color level at 5 minutes, leave the highlight on another minute, up to 7 minutes *total* work time.

10. Apply the triple-barrier cream and put on gloves.

11. Shake the mousse and apply it from the scalp to the ends. Comb through to disperse thoroughly.

12. As soon as you are done with the application, saturate cotton balls with aerosol hair spray as needed and wipe away the triple-barrier cream in the direction of the hairline to avoid creating a harsh line. This ensures you will not stain your skin.

13. Set the timer for 15 minutes.

14. When time is up, rinse your hair and shampoo lightly with a color-friendly shampoo. Condition, detangle, and style as usual.

15. Maintain your new color by using the L'Oréal Color Pulse Mousse in Chilled Plum once a week.

Tabs 4 through 11

Tabs 4 through 11 must use your tab's recipe for Milk Chocolate (Chapter 5, page 100) to get the brown you need *before* continuing on with the highlight and Chilled Plum Pulse Mousse. After completing the process for Milk Chocolate, you will proceed with the exact highlight procedure and mousse treatment for Tabs 1 through 3 (page 44).

MERLOT

Preparing this color is almost as easy as popping a cork! And the color is exciting and very mysterious. It is important to use Garnier Nutrisse colors for this recipe. They have a very specific tone that we need to get just right!

Tabs 1 and 2

I do not recommend Merlot for Tabs 1 and 2.

Tabs 3 and 4

MATERIALS

Triple-barrier cream

1 box light blonde

8 cotton balls

Aerosol hair spray

Timer

Garnier Nutrisse Medium Reddish Brown #56

Shampoo and conditioner

L'Oréal Paris Colorist Collection Mahogany Shampoo*

*Available at beauty supply stores and online (see the Resources section on page 289)

TRIPLE-BARRIER CREAM—A REMINDER

First, with your fingertips apply a layer of lotion, such as Nivea or Keri, around your face at your hairline. Follow with a layer of petroleum jelly. Then along the outside border of the lotion/petroleum jelly strip, apply a layer of Carmex lip balm.

---

1. Detangle your hair and comb it away from your face.

2. Apply the triple-barrier cream and put on gloves.

3. Prepare the light blonde color according to the package directions.

4. Apply the mixture to your hair, starting 1 inch from your scalp down through the ends. Comb through your hair to ensure the mixture is distributed evenly.

5. Set the timer for 10 minutes.

6. When time is up, rinse your hair, shampoo gently with no scalp manipulation, condition lightly, towel blot, and detangle.

7. Apply the triple-barrier cream and put on gloves.

8. Prepare Garnier Nutrisse Medium Reddish Brown according to the package instructions.

9. Apply the mixture to your hair, starting from your scalp through to the ends. Comb through your hair to ensure the mixture is distributed evenly.

10. As soon as you are done with the application, saturate cotton balls with aerosol hair spray as needed and wipe away the triple-barrier cream in the direction of the hairline to avoid creating a harsh line. This ensures you will not stain your skin.

11. Set the timer for 20 minutes.

12. When time is up, rinse your hair, shampoo gently with no scalp manipulation, condition lightly, and towel dry. Style as usual.

13. Every third shampoo, start with a cleansing shampoo and do a second shampoo using Colorist Collection Mahogany Shampoo. It is necessary in order to maintain the richness of this color.

Tab 5

Triple-barrier cream

1 box light blonde

Timer

Shampoo and conditioner

1 box Garnier Nutrisse Medium Reddish Brown #56

Measuring cups and spoons

1 box Garnier Nutrisse Rich Auburn Blonde #76

8 cotton balls

Aerosol hair spray

L'Oréal Paris Colorist Collection Mahogany Shampoo*

*Available at beauty supply stores and online (see the Resources section on page 289)

DIRECTIONS

1. Detangle your hair and comb it away from your face.

2. Apply the triple-barrier cream.

3. Prepare the light blonde mixture according to the package directions.

4. Apply the light blonde mixture, starting 1 inch from the scalp down through the ends. Comb through to ensure even coverage.

5. Set the timer for 5 minutes.

6. When time is up, rinse your hair, shampoo gently with no scalp manipulation, condition lightly, detangle, and towel blot well.

7. Open the developer from the package of the Garnier Nutrisse Medium Reddish Brown. Using measuring cups or spoons, add 1 ounce (½ color container) of the medium reddish brown tint.

8. Add 1 ounce (½ color container) of the Rich Auburn Blonde.

9. Shake the bottle to mix well, holding your gloved finger over the open tip.

10. Apply to your hair starting from your scalp down through to the ends. Comb through your hair to ensure mixture is distributed evenly.

11. As soon as you are done with the application, saturate cotton balls with aerosol hair spray as needed and wipe away the triple-barrier cream in the direction of the hairline to avoid creating a harsh line. This ensures you will not stain your skin.

12. Set timer for 30 minutes.

13. When time is up, rinse your hair and shampoo gently. Use the Colorist Collection Mahogany Shampoo for a second shampoo. Condition, towel blot, and style as usual.

14. Use the Colorist Collection Mahogany every third shampoo after that.

Tabs 5½ and 6

MATERIALS

Triple-barrier cream

1 box Garnier Nutrisse Medium Reddish Brown #56

Measuring cups and spoons

1 box Garnier Nutrisse Medium Reddish Rich Auburn Blonde #76

8 cotton balls

Aerosol hair spray

Timer

Shampoo and conditioner

DIRECTIONS

1. Detangle your hair and comb it away from your face.

2. Apply the triple-barrier cream and put on gloves.

3. Remove the developer from the Garnier Nutrisse Medium Reddish Brown kit, snip off the applicator tip, and remove the cap.

4. Using measuring cups or spoons, add 1 ounce (½ color container) of the Garnier Nutrisse Medium Reddish Brown tint to the developer.

5. Add 1 ounce (½ color container) of the Rich Auburn Blonde tint to the developer.

6. Shake the bottle to mix the materials, holding your gloved finger over the open tip.

7. Apply the mixture to your hair, starting at 1 inch from the scalp down through to the ends. Comb through your hair to make sure the mixture is distributed evenly.

9. As soon as you are done with the application, saturate cotton balls with aerosol hair spray as needed and wipe away the triple-barrier cream in the direction of the hairline to avoid creating a harsh line. This ensures you will not stain your skin.

10. Set the timer for 20 minutes.

11. When time is up, rinse your hair, shampoo well, and use Colorist Collection in Mahogany as a second shampoo. Condition, towel dry, and style as usual. Use the Colorist Collection every third shampoo after that.

Tab 6½

Triple-barrier cream

1 box reddish brown (I prefer Garnier Nutrisse #56)

8 cotton balls

Aerosol hair spray

Timer

Shampoo and conditioner

DIRECTIONS

1. Detangle hair and comb it away from your face.

2. Apply the triple-barrier cream and put on gloves.

3. Mix the reddish brown kit according to the manufacturer's instructions.

4. Shake the bottle gently to mix the materials, holding your gloved finger over the open tip.

5. Apply the mixture to your hair starting from your scalp through to the ends. Comb through to ensure even distribution.

6. As soon as you are done with the application, saturate cotton balls with aerosol hair spray as needed and wipe away the triple-barrier cream in the direction of the hairline to avoid creating a harsh line. This ensures you will not stain your skin.

7. Set the timer for 20 minutes.

8. When time is up, rinse your hair, shampoo gently with no scalp manipulation, condition lightly, and towel dry. Style as usual.

MATERIALS

Triple-barrier cream

1 box Garnier Nutrisse Medium Reddish Brown #56

Measuring cups and spoons

1 box dark golden brown

8 cotton balls

Aerosol hair spray

Timer

Shampoo and conditioner

L'Oréal Paris Colorist Collection Mahogany Shampoo*

*Available at beauty supply stores and online (see the Resources section on page 289)

DIRECTIONS

1. Detangle your hair and comb it away from your face.

2. Apply the triple-barrier cream and put on gloves.

3. Remove the developer from the medium reddish brown kit, snip off the applicator tip, and remove the cap.

4. Using measuring cups or spoons, add 1¾ ounces (three ¼ + ⅛ color container) of the Garnier Nutrisse Medium Reddish Brown tint to the developer.

5. Add ¼ ounce (⅛ color container) dark golden brown tint to the developer.

7. Gently shake the bottle to mix the materials, holding your gloved finger over the open tip.

8. Apply the mixture to your hair, starting from your scalp through to the ends. Comb through to ensure even distribution.

9. As soon as you are done with the application, saturate cotton balls with aerosol hair spray as needed and wipe away the triple-barrier cream in the direction of the hairline to avoid creating a harsh line. This ensures you will not stain your skin.

10. Set the timer for 30 minutes.

11. When time is up, rinse your hair, shampoo well, use Colorist Collection Mahogany as a second shampoo, condition, and towel dry. Style as usual. Use the Mahogany every third shampoo to prevent fading.

Tab 8

MATERIALS

Triple-barrier cream

1 box medium reddish brown (I like Garnier Nutrisse #56)

Measuring cups and spoons

1 box dark golden brown

¼ teaspoon dark auburn

Aerosol hair spray

Timer

Shampoo and conditioner

DIRECTIONS

1. Detangle your hair and comb it away from your face.

2. Apply the triple-barrier cream and put on gloves.

3. Remove the developer from the medium reddish brown, snip off the applicator tip, and remove the cap.

4. Using measuring cups or spoons, add 1½ ounces (three ¼ color container) of the medium reddish brown tint to the developer.

5. Add ½ ounce (¼ color container) of the dark golden brown tint to the developer.

6. Add ¼ teaspoon of dark auburn tint to the developer.

7. Shake the bottle to mix the materials, holding your gloved finger over the open tip.

8. Apply the mixture to your hair, starting from your scalp through to the ends. Comb through to ensure even coverage.

9. As soon as you are done with the application, saturate cotton balls with aerosol hair spray as needed and wipe away the triple-barrier cream in the direction of the hairline to avoid creating a harsh line. This ensures you will not stain your skin.

10. Set the timer for 20 minutes.

11. When time is up, rinse your hair, shampoo gently with no scalp manipulation, gently condition, and towel dry. Style as usual.

Tabs 9 and 10

Revert to the recipe for your pre-gray natural hair-color tab.

Tab 11

MATERIALS

1 perm kit

Triple-barrier cream

1 box dark golden brown

Measuring cups and spoons

1 box medium reddish brown (I prefer Garnier Nutrisse #56)

8 cotton balls

Aerosol hair spray

Timer

Shampoo and conditioner

L'Oréal Paris Colorist Collection Red Mahogany Shampoo*

Available at beauty supply stores and online (see the Resources section on page 289)

DIRECTIONS

1. Pre-soften hair: Apply the perm-kit solution to your hair starting about 1¼ inches from your scalp down to the ends and let it work for 1 minute. Rinse, shampoo with no scalp manipulation, and comb out your hair. You are ready to continue with your recipe.

2. Using measuring cups or spoons, add ¾ ounce (¼ + ⅛ color container) of dark golden brown tint to the developer.

3. Add 1¼ ounces (½ + ⅛ color container) of medium reddish brown to the tint.

4. Shake to mix the materials, holding your gloved finger over the open tip.

5. Apply the mixture to your hair starting from your scalp through to the ends. Comb through your hair to make sure the mixture is distributed evenly.

6. As soon as you are done with the application, saturate cotton balls with aerosol hair spray as needed and wipe away the triple-barrier cream in the direction of the hairline to avoid creating a harsh line. This ensures you will not stain your skin.

7. Set the timer for 25 minutes.

8. When time is up, rinse your hair and shampoo well. Shampoo a second time with Colorist Collection in Mahogany, condition lightly, and towel dry. Style as usual. Use the Mahogany every third shampoo to maintain depth.

SPICED PERSIMMON

This red is provocative—a real mouthwatering look—so be prepared to hear fire engines blaring their horns when you walk down the street! One note to Spiced Persimmon ladies: You must use the appropriate Color Pulse mousse once a week. Follow the product instructions for application.

Tabs 1 through 5

I suggest this recipe for girls with short, virgin hair, because bleach does take its toll!

MATERIALS

Triple-barrier cream

1 box Clairol Born Blonde Maxi Bleach

L'Oréal Paris Color Pulse Mousse in Red Copper Spritz and Iced Coffee

Plastic bowl

Q-tip

8 cotton balls

Aerosol hair spray

Timer

Shampoo and conditioner

DIRECTIONS

1. Detangle your hair and comb it away from your face.

2. Apply the triple-barrier cream and put on gloves.

3. Open the Clairol Maxi Bleach kit and mix it according to the package directions.

4. Apply the mixture to your hair, starting 1 inch down from your scalp through to the ends. Comb through carefully for even coverage.

5. Set the timer for 10 minutes. Apply the mixture 1 inch from your scalp, checking carefully for missed spots!

6. Your goal color here is a dark caramel—refer to the instructions on the package.

7. When time is up and the correct color is achieved, rinse your hair, shampoo gently, and detangle.

8. Reapply the triple-barrier cream and put on gloves.

9. In a plastic bowl, combine ½ cup of the Red Copper Spritz Mousse and ½ cup of the Iced Coffee Mousse. Use a Q-tip to mix well.

10. Apply the mixture to your hair starting from the scalp through to the ends. Comb through your hair to make sure the mixture is distributed evenly.

11. Set the timer for 15 minutes.

12. As soon as you are done with the application, saturate cotton balls with aerosol hair spray as needed and wipe away the triple-barrier cream in the direction of the hairline to avoid creating a harsh line. This ensures you will not stain your skin.

13. When time is up, check your color. If you would like a redder shade, apply the Red Copper Spritz Mousse for another 5 minutes. You can always adjust the proportion and timing the next time.

14. Rinse your hair and shampoo gently with a color-friendly shampoo, condition, and towel dry. Style as usual.

Tabs 5½ and 6

MATERIALS

Triple-barrier cream

1 box light blonde

1 box L'Oréal Paris Red Copper Spritz Color Pulse Mousse

8 cotton balls

Aerosol hair spray

Timer

Shampoo and conditioner

DIRECTIONS

1. Detangle your hair and comb it away from your face.

2. Apply the triple-barrier cream and put on gloves.

3. Prepare the light blonde tint according to the package instructions. Apply the mixture to your hair, starting ½ inch from your scalp down through to the ends. Comb through your hair to make sure the mixture is distributed evenly. Then apply the mixture at the scalp to ½ inch down.

4. Set the timer for 3 minutes.

5. When time is up, rinse your hair and shampoo gently with no scalp manipulation, condition lightly, and towel dry.

6. Open the Red Copper Spritz Mousse and shake well.

7. Apply triple-barrier cream and put on gloves.

8. Apply the mousse from the scalp through to the ends, combing through to ensure even coverage.

9. Set the timer for 10 minutes.

10. When time is up, rinse, shampoo with a color-friendly shampoo, condition, and style as usual.

Tab 6½

1 box light blonde (I like Revlon High Dimension or L'Oréal Paris Féria)

Plastic bowl

Tung brush

Homemade water-jug paddle

Timer

Shampoo and conditioner

Triple-barrier cream

L'Oréal Paris Red Copper Spritz Color Pulse Mousse

DIRECTIONS

1. Detangle your hair and comb it away from your face.

2. Mix the light blonde tint according to the manufacturer's instructions and put it in a plastic bowl.

3. Dip the Tung brush into the blonde mixture and apply boldly to your hair from ½ inch from your part through to the ends. (Use the ribboning technique on page 47.) Be sure to use your water-jug paddle, and keep the application bottom heavy.

4. Set the timer for 10 minutes.

5. When time is up, rinse your hair, shampoo gently with no scalp manipulation, and condition lightly. Detangle. Towel blot well.

6. Apply the triple-barrier cream.

7. Prepare L'Oréal Red Copper Spritz Mousse according to the package instructions and put on gloves. Apply the mousse to your hair starting from your scalp through to the ends. Comb through to ensure even coverage.

8. Set the timer for 7 minutes.

9. When time is up, rinse your hair, shampoo well with a color-friendly shampoo, condition lightly, and towel dry. Style as usual.

10. Use the mousse once a week for maximum vibrancy.

Tabs 7 and 7½

MATERIALS

Triple-barrier cream

1 box reddish blonde or strawberry blonde

Timer

8 cotton balls

Aerosol hair spray

Shampoo and conditioner

L'Oréal Paris Red Copper Spritz Color Pulse Mousse

L'Oréal Paris Iced Coffee Color Pulse Mousse

Measuring cups and spoons

Plastic bowl

Q-tips

DIRECTIONS

1. Detangle your hair and comb it away from your face.

2. Apply the triple-barrier cream and put on gloves.

3. Take the developer out of the kit, snip off the applicator tip, and remove the cap.

4. Open the box of color and mix according to the manufacturer's instructions.

5. Apply the mixture to your hair starting from 1 inch from your scalp through to the ends. Comb through to ensure even coverage.

6. Set the timer for 5 minutes.

7. When time is up, apply the tint from the scalp to the 1-inch mark. Comb through to ensure even coverage.

8. As soon as you are done with the application, saturate cotton balls with aerosol hair spray as needed and wipe away the triple-barrier cream in the direction of the hairline to avoid creating a harsh line. This ensures you will not stain your skin.

9. Set the timer for 20 minutes.

10. When time is up, rinse your hair, shampoo well, condition lightly, and towel blot. Detangle your hair.

11. Open both Color Pulse mousses and shake well. Put on gloves.

12. Measure ½ cup of each mousse color into the plastic bowl. Mix well (about 10 to 15 seconds) with a Q-tip.

13. Use your gloved hands to spread the mousse mixture from your scalp through to the ends. Comb through carefully.

14. Set the timer for 7 minutes.

15. When time is up, rinse your hair, shampoo with a color-friendly shampoo, and condition lightly. Towel dry and style as usual.

16. Use the mousse mixture once a week to ensure vibrant color.

Tab 8

Tab 8

MATERIALS

Triple-barrier cream

1 box light auburn (I like Nice'n Easy Light Auburn #10)

8 cotton balls

Aerosol hair spray

Timer

Shampoo and conditioner

L'Oréal Paris Iced Coffee Color Pulse Mousse

L'Oréal Paris Red Copper Spritz Color Pulse Mousse

Measuring cups and spoons

Plastic bowl

Q-tips

DIRECTIONS

1. Detangle your hair and comb it away from your face.

2. Apply the triple-barrier cream.

3. Take the developer out of the kit, snip off the applicator tip, and remove the cap.

4. Mix the color according to the manufacturer's instructions.

5. Replace the cap and shake to mix the materials, holding your gloved finger over the open tip.

6. Apply the mixture to your hair starting from your scalp through to the ends.

7. As soon as you are done with the application, saturate cotton balls with aerosol hair spray as needed and wipe away the triple-barrier cream in the direction of the hairline to avoid creating a harsh line. This ensures you will not stain your skin.

8. Set the timer for 30 minutes.

9. When time is up, rinse your hair, shampoo gently with no scalp manipulation, condition lightly, and towel blot. Detangle your hair.

10. Reapply the triple-barrier cream.

11. Open both packages of mousse and shake the bottles well. Put on gloves. Using measuring cups, place in a plastic bowl ¾ cup of the Iced Coffee Mousse and ¼ cup of the Red Copper Spritz Mousse.

12. Stir well with a Q-tip until combined, about 15 seconds.

13. Use your gloved hands to spread the mousse mixture from your scalp through to the ends. Comb through to ensure even coverage.

14. Set the timer for 10 minutes.

15. When time is up, rinse your hair, shampoo gently, and condition lightly. Towel dry and style as usual.

16. Use the Color Pulse mousse mixture once a week to keep your red vibrant.

Tabs 9 and 10

Revert to your pre-gray natural color tab recipe.

Tab 11

MATERIALS

1 perm kit

Triple-barrier cream

1 box light auburn (I like Clairol Nice'n Easy #110)

1 box light golden brown

Measuring cups and spoons

8 cotton balls

Aerosol hair spray

Timer

Shampoo and conditioner

L'Oréal Paris Iced Coffee Color Pulse Mousse

L'Oréal Paris Red Copper Spritz Color Pulse Mousse

Plastic Bowl

Q-tips

DIRECTIONS

1. Pre-soften your hair: apply the solution to your hair starting about 1¼ inches from your scalp down to the ends and let it work for 1 minute. Rinse, shampoo with no scalp manipulation, and comb out your hair. You are ready to continue with your recipe.

2. Apply the triple-barrier cream and put on gloves.

3. Take the developer out of the light auburn kit, snip off the applicator tip, and remove the cap.

4. Using measuring cups or spoons, add 1½ ounces (¾ color container) of light auburn tint to the developer.

5. Add ½ ounce (¼ color container) of the light golden brown to the developer.

6. Shake the bottle to mix the materials, holding your gloved finger over the open tip.

7. Apply the mixture to your hair starting from your scalp through to the ends. Comb through to ensure even coverage.

8. As soon as you are done with the application, saturate cotton balls with aerosol hair spray as needed and wipe away the triple-barrier cream in the direction

of the hairline to avoid creating a harsh line. This ensures you will not stain your skin.

9. Set the timer for 30 minutes.

10. When time is up, rinse your hair, shampoo gently, condition lightly, and towel blot. Detangle your hair.

11. Reapply the triple-barrier cream.

12. Open both packages of mousse and shake the bottles well. Put on gloves. Using measuring cups, place in a plastic bowl ¾ cup of the Iced Coffee mousse and ¼ cup of the Red Copper Spritz mousse.

13. Stir well with a Q-tip until combined, about 15 seconds.

14. Use your gloved hands to spread the mousse mixture from your scalp through to the ends. Comb through to ensure even coverage.

15. Set the timer for 10 minutes.

16. When time is up, rinse your hair, shampoo with a color-friendly shampoo, and condition lightly. Towel dry and style as usual.

17. Use the mousse mixture once a week to ensure vibrant color.

NOTE: If you want a richer result, next time mix ⅛ ounce (¹⁄₁₆ color container) of medium golden brown and up to 1 ounce (½ color container) of light golden brown and proceed as directed in the recipe above.

GINGERED TOFFEE

This is a sophisticated, soft red, perfect for those of you who want a subtler look.

NOTE: All ginger-haired girls should use L'Oréal Paris Colorist Collection Ginger Root Shampoo every third shampoo.

Tabs 1, 2, 3, and 4

Gingered Toffee is not applicable to the darker Tabs 1 through 4.

Tabs 5

MATERIALS

1 box light blonde

L'Oréal Paris Couleur Experte 6.4

8 cotton balls

Aerosol hair spray

Timer

Shampoo and conditioner

Triple-barrier cream

L'Oréal Paris Colorist Collection Ginger Root Shampoo

Available at beauty supply stores and online (see the Resources section on page 289)

DIRECTIONS

1. Detangle your hair and comb it away from your face.

2. Prepare the light blonde tint according to the package instructions.

3. Apply the mixture to your hair starting from 1½ inches down from your scalp through to the ends. Comb through your hair to make sure the mixture is distributed evenly.

4. Set the timer for 5 minutes.

5. When the time is up, apply the remainder of the mixture starting from your scalp through to 1½ inches down, and leave it on for the remaining 2 minutes. Comb through your hair to make sure the mixture is distributed evenly.

6. When time is up, rinse your hair, shampoo gently, and towel blot well. Detangle your hair.

7. Apply the triple-barrier cream and put on gloves.

8. Prepare L'Oréal Couleur Experte 6.4 according to the manufacturer's instructions.

9. Apply the mixture starting from your scalp down through the ends. Comb through your hair to make sure the mixture is distributed evenly.

10. As soon as you are done with the application, saturate cotton balls with aerosol hair spray as needed and wipe away the triple-barrier cream in the direction of the hairline to avoid creating a harsh line. This ensures you will not stain your skin.

11. Set the timer for 25 minutes.

12. When time is up, rinse your hair, shampoo gently, condition lightly, and towel dry.

13. Use the Couleur Experte bleach packet as directed to add beautiful dimension to your hair. I suggest the TTR technique found on page 48.

14. Follow with Colorist Collection Ginger Root Shampoo, condition, towel blot, and style as usual.

Tabs 5½ and 6

MATERIALS

Triple-barrier cream

1 box light reddish blonde

8 cotton balls

Aerosol hair spray

Timer

Shampoo and conditioner

Standard highlight recipe (see page 44)

L'Oréal Paris Colorist Collection Ginger Root Shampoo

Available at beauty supply stores and online (see the Resources section on page 289)

DIRECTIONS

1. Detangle your hair and comb it away from your face.

2. Apply the triple-barrier cream and put on gloves.

3. Prepare the light reddish blonde tint according to the package instructions.

4. Apply the mixture to your hair starting from 1½ inches down from your the scalp through to the ends. Comb through your hair to make sure the mixture is distributed evenly.

5. As soon as you are done with the application, saturate cotton balls with aerosol hair spray as needed and wipe away the triple-barrier cream in the direction of the hairline to avoid creating a harsh line. This ensures you will not stain your skin.

6. Set the timer for 30 minutes.

7. When time is up, rinse your hair, shampoo gently, and towel blot. Detangle your hair and part it your favorite way.

8. Prepare the standard highlight recipe (page 44) and highlight your hair using the TTR technique (page 48).

9. Check the highlight after about 15 minutes. You're looking to achieve a light caramel color.

10. Rinse your hair, shampoo, and shampoo again using Colorist Collection Ginger Root Shampoo. Condition lightly, towel dry, and style as usual.

Tab 6½

Since your hair is very close to this color, I recommend using the TTR technique (page 48) with the standard highlight recipe (page 44). Check the highlight after about 15 minutes. You're looking to achieve a light caramel color. Follow up with Colorist Collection Ginger Root Shampoo.

Tabs 7 and 7½

MATERIALS

Triple-barrier cream

1 box L'Oréal Paris Couleur Experte Light Golden Copper Brown

8 cotton balls

Aerosol hair spray

Timer

Shampoo and conditioner

L'Oréal Paris Colorist Collection Ginger Root Shampoo*

Available at beauty supply stores and online (see the Resources section on page 289)

DIRECTIONS

1. Detangle your hair and comb it away from your face.

2. Apply the triple-barrier cream and put on gloves.

3. Take the developer out of the Light Golden Copper Brown box, snip off the applicator tip, and remove the cap.

4. Add the Light Golden Copper Brown tint to the developer.

5. Shake the bottle to mix the materials, holding your gloved finger over the open tip.

6. Apply the mixture to your hair starting from your scalp through to the ends. Comb through your hair to make sure the mixture is distributed evenly.

7. As soon as you are done with the application, saturate cotton balls with aerosol hair spray as needed and wipe away the triple-barrier cream in the direction of the hairline to avoid creating a harsh line. This ensures you will not stain your skin.

8. Set the timer for 30 minutes.

9. When time is up, rinse your hair, shampoo well, condition gently, and towel dry. Style as usual.

10. Highlight your hair using the bleach found in your Couleur Experte kit. I recommend following the TTR technique on page 48.

11. Follow with Colorist Collection Ginger Root Shampoo. Condition, towel blot, and style as usual.

Tab 8

MATERIALS

Triple-barrier cream

1 box L'Oréal Paris Couleur Experte Light Golden Copper Brown

8 cotton balls

Aerosol hair spray

Timer

Shampoo and conditioner

L'Oréal Paris Colorist Collection Ginger Root Shampoo*

Available at beauty supply stores and online (see the Resources section on page 289)

1. Detangle your hair and comb it away from your face.

2. Apply the triple-barrier cream and put on gloves.

3. Take the developer out of the Light Golden Copper Brown box, snip off the applicator tip, and remove the cap.

4. Add the Light Golden Copper Brown tint to the developer.

5. Shake the bottle to mix the materials, holding your gloved finger over the open tip.

6. Apply the mixture to your hair starting from your scalp through to the ends. Comb through your hair to make sure the mixture is distributed evenly.

7. As soon as you are done with the application, saturate cotton balls with aerosol hair spray as needed and wipe away the triple-barrier cream in the direction of the hairline to avoid creating a harsh line. This ensures you will not stain your skin.

8. Set the timer for 30 minutes.

9. When time is up, rinse your hair, shampoo gently with no scalp manipulation, condition gently, and towel dry. Style as usual.

10. Prepare the highlight kit found in your Couleur Experte kit and highlight your hair following the TTR technique (see page 48).

11. Leave the bleach in for about 15 to 20 minutes, until you've reached a light caramel color.

12. Follow with Colorist Collection Ginger Root Shampoo. Condition, towel blot, and style as usual.

Tab 9

Revert to your pre-gray tab recipe.

Tab 10

If you're closer to 35 percent gray, revert to your pre-gray tab recipe. If you're closer to 65 percent gray, follow the recipe for Tab 11.

Tab 11

MATERIALS

1 perm kit

Triple-barrier cream

1 box L'Oréal Paris Couleur Experte Light Golden Copper Brown

Measuring cups and spoons

1 box L'Oréal Paris Féria Deep Copper

8 cotton balls

Aerosol hair spray

Timer

Shampoo and conditioner

L'Oréal Paris Colorist Collection Ginger Root Shampoo*

Available at beauty supply stores and online (see the Resources section on page 289)

DIRECTIONS

1. Pre-soften your hair. Apply the permanent solution from the drugstore-perm kit to your hair starting about 1¼ inches from your scalp down to the ends and let it work for 1 minute. Rinse, shampoo, and comb out your hair. You are ready to continue with your recipe.

2. Apply the triple-barrier cream and put on gloves.

3. Take the developer out of the Light Golden Copper Brown box, snip off the applicator tip, and remove the cap.

4. Add the Light Golden Copper Brown tint to the developer.

5. Add 2 teaspoons of the Deep Copper tint to developer.

6. Shake the bottle to mix the materials, holding your gloved finger over the open tip.

7. Apply the mixture to your hair starting from your scalp through to the ends. Comb through your hair to make sure the mixture is distributed evenly.

8. As soon as you are done with the application, saturate cotton balls with aerosol hair spray as needed and wipe away the triple-barrier cream in the direction of the hairline to avoid creating a harsh line. This ensures you will not stain your skin.

9. Set the timer for 30 minutes.

10. When time is up, rinse your hair, shampoo gently with no scalp manipulation, condition lightly, and towel dry. Style as usual.

11. Take the developer out of the Deep Copper box, snip off the applicator tip, and remove the cap.

12. Remove 2 teaspoons of the developer and add the remaining Deep Copper tint to the developer bottle.

13. Shake the bottle to mix the materials, holding your gloved finger over the open tip.

14. Prepare the Couleur Experte highlight mixture according to the manufacturer's directions.

15. Using the Peekaboo Punk technique (page 236), apply three 1-inch ribbons with the Deep Copper tint.

16. Using the TTR technique (page 48), apply the highlight mixture.

17. Set the timer for 15 minutes.

18. When time is up, check the highlight color. You are looking to achieve a light caramel color.

19. Remove the foil from the Deep Copper tint.

20. Rinse well, shampoo, and do a second shampoo using Colorist Collection Ginger Root Shampoo. Condition, towel blot, and style as usual.

APRICOT

This delightfully clear, bright hue is cheerful and warm on lighter complexions with warm eye colors, and it feels sexy and bold on darker complexions. When applying makeup, you can complement your new hair color with a little extra peach to tawny blush, depending on your skin tone. It's perfect!

Tab 1

I prefer this recipe not be used on Tabs 1 to 5, *unless* you have very short hair—well above your chin. It is a two-step process, and it can be very hard on darker hair. If you have long hair and you want this color, you must see a salon professional. In addition, if you have had your hair straightened or permed but have never colored your hair, you *must not* do this at home unless it is a touch-up on regrowth.

Tabs 1 through 5 (Short Hair)

For those of you with very short, virgin hair, this recipe can work.

MATERIALS

Triple-barrier cream

1 box Clairol Born Blonde Maxi Bleach

8 cotton balls

Aerosol hair spray

Timer

1 pink grapefruit (to match color)

Shampoo and conditioner

L'Oréal Paris Colorist Collection Ginger Root Shampoo*

*Available at beauty supply stores and online (see the Resources section on page 289)

1. Detangle your hair and comb it away from your face.

2. Apply the triple-barrier cream and put on gloves.

3. Apply the Maxi Bleach according to the package instructions.

4. Apply approximately ¾ the mixture to your hair starting ½ inch down from your scalp through to the ends. Comb through your hair to make sure the mixture is distributed evenly.

5. Set the timer for 30 minutes.

6. After 15 minutes, apply the remainder of the mixture starting from your scalp to 1½ inches down and leave it in for the remaining time. Comb through your hair to make sure the mixture is distributed evenly.

7. Follow the manufacturer's directions and check the strands about every 5 minutes. Your color is ready when the hair matches the most yellow part of the outside of a pink grapefruit rind.

8. Rinse your hair, shampoo it gently with no scalp manipulation, and shampoo again using Colorist Collection Ginger Root Shampoo. Condition well, towel blot, and style as usual.

Tabs 5½, 6, and 6½

I strongly urge you to consider using this recipe as a highlight only, but if you must have an all-over application, proceed with the following recipe.

MATERIALS

1 box light blonde

Timer

Shampoo and conditioner

L'Oréal Paris Colorist Collection Ginger Root Shampoo*

*Available at beauty supply stores and online (see the Resources section on page 289)

1. Prepare the light blonde tint according to the package instructions. Apply ¾ the mixture to your hair starting 1½ inches from your scalp through to your ends.

2. Set timer for 5 minutes.

3. When time is up, apply the remaining mixture to your hair starting from the scalp down to the previous application.

4. Set the timer for 5 minutes.

5. When time is up, rinse your hair, shampoo, and shampoo a second time using Colorist Collection Ginger Root Shampoo. Condition lightly, towel blot, and style as usual.

Tabs 7 and 7½

MATERIALS

1 box light golden blonde

Timer

Shampoo and conditioner

L'Oréal Paris Colorist Collection Ginger Root Shampoo*

*Available at beauty supply stores and online (see the Resources section on page 289)

DIRECTIONS

1. Detangle your hair and comb it away from your face.

2. Take the developer out of the light golden blonde box, snip off the applicator tip, and remove the cap.

3. Add the light golden blonde tint to the developer.

4. Shake the bottle to mix the materials, holding your gloved finger over the open tip.

5. Apply approximately ¾ the mixture to your hair starting 1 inch down from your scalp through to the ends. Comb through your hair to make sure the mixture is distributed evenly.

6. Let it work for 10 minutes, then apply the remainder of the mixture starting from your scalp through 1 inch down. Comb through your hair to make sure the mixture is distributed evenly.

7. Set the timer for 10 minutes.

8. When time is up, rinse your hair, shampoo, and shampoo again using Colorist Collection Ginger Root Shampoo. Condition, towel blot, and style as usual. Use the Ginger Root Shampoo every third shampoo to help maintain your color.

Tab 8

MATERIALS

1 box light golden blonde

Timer

Shampoo and conditioner

L'Oréal Paris Colorist Collection Ginger Root Shampoo*

Available at beauty supply stores and online (see the Resources section on page 289)

DIRECTIONS

1. Detangle your hair and comb it away from your face.

2. Prepare the light golden blonde tint according to the package directions.

3. Apply approximately ¾ the mixture to your hair starting 1 inch from your scalp down to the ends. Comb through your hair to make sure the mixture is distributed evenly.

4. Set the timer for 5 minutes.

5. After 3 minutes, apply the remainder of the mixture starting from your scalp to 1 inch down and leave it in for the remaining 2 minutes.

6. When time is up, rinse your hair, shampoo it with a cleansing shampoo, and shampoo again using Colorist Collection Ginger Root Shampoo. Condition lightly, towel blot, and style as usual.

Tab 9

Revert to your pre-gray natural hair-color tab recipe.

Tab 10

If you're closer to 35 percent gray, revert to your pre-gray natural hair-color tab recipe. If it is closer to 65 percent gray, use the recipe for Tab 11.

Tab 11

MATERIALS

1 perm kit

Triple-barrier cream

1 box light reddish blonde

Measuring cups and spoons

1 box medium golden blonde

8 cotton balls

Aerosol hair spray

Timer

Shampoo and conditioner

L'Oréal Paris Colorist Collection Ginger Root Shampoo*

Available at beauty supply stores and online (see the Resources section on page 289)

1. Pre-soften your hair. Apply the perm-kit solution to your hair starting about 1¼ inches from your scalp down to the ends and let it work for 1 minute. Rinse, shampoo with no scalp manipulation, and comb out your hair. You are ready to continue with your recipe.

2. Apply the triple-barrier cream and put on gloves.

3. Take the developer out of the light reddish blonde box, snip off the applicator tip, and remove the cap.

4. Using measuring cups or spoons, add 1 ounce (½ color container) of the light reddish blonde tint to the developer.

5. Add 1 ounce (½ color container) of the medium golden blonde tint to the developer.

6. Shake the bottle to mix the materials, holding your gloved finger over the open tip.

7. Apply the mixture to your hair starting from the scalp through to the ends. Comb through your hair to make sure the mixture is distributed evenly.

8. As soon as you are done with the application, saturate cotton balls with aerosol hair spray as needed and wipe away the triple-barrier cream in the direction of the hairline to avoid creating a harsh line. This ensures you will not stain your skin.

9. Set the timer for 30 minutes.

10. When time is up, rinse your hair, shampoo, and shampoo again using Colorist Collection Ginger Root Shampoo. Condition, towel dry, and style as usual.

NOTE: If you find you want more coverage next time, try adding a cap of light golden brown to the developer.

8.

SCRUMPTIOUS, SUMPTUOUS BLONDES

This chapter helps naturally fair-haired beauties let the sun shine indoors and out, night and day. It's important to remember that natural-looking blonde is a difficult color to achieve if you are very dark-haired. If you are a Tab 1 through 5 and you are determined to be blonde, go immediately to a salon. Do not try this at home—you *will* be disappointed. Even a salon expert may not be able to get you where you want to go. Or, you can let some light in with highlights that are outlined in Chapters 6 and 7 or have a salon give you lighter highlights if you are a true dark brunette or black.

Blonde also demands skin-and eye-color compatibility; to match a naturally darker complexion with blonde hair, you may have to do plenty of extra work in terms of makeup and maybe even an eye-color change via cosmetic contact lenses. Is this work worth it to you?

Tabs 1 through 5 and Tabs 10 and 11 (unless you are naturally a Tab 6 through 9) should also bypass Vanilla'd Butterscotch and Brandied Fig. These are over-the-top blondes, and it is necessary to have a natural blonde base with which to work. Blonde is a delicate color, and while it is imperative that all recipes in this book are followed closely the first time you do them, it is particularly important to adhere strictly to the recipes in this chapter.

Our flaxen target colors are:

- **AMBER HONEY DREAM WITH LEMON BLOSSOM RIBBONS**

- **VANILLA DREAMSICLE CREAMSICLE**

- **VANILLA'D BUTTERSCOTCH**

- **BRANDIED FIG**

NOTE: If you have very thick, long hair (shoulder length or longer), double all recipes.

PREPARATION IS EVERYTHING

Before you begin any recipe, read through the ingredient list, get out your color kit, and prepare your work area. Clear off the countertop on which you will be working and, spread out your materials. Detangle your hair without manipulating your scalp. Put on your work shirt, and wrap a towel around your neck to catch stains. Apply a barrier of Vaseline brand petroleum jelly around your hairline. Blonde tints do not stain as badly as darker hues, but that doesn't mean you shouldn't protect the skin at your hairline. Cover the floor with a drop cloth or old newspapers. (Remember, no plastic underfoot!) Be sure to have your timer out and ready to set. *Wear the gloves that came in your color kit before beginning any coloring procedure!* You can recycle gloves by rinsing them inside and out, patting them dry, and sprinkling some baby powder or cornstarch on the inside for easy slip-on next time.

AMBER HONEY DREAM
WITH LEMON BLOSSOM RIBBONS

As I mentioned in the introduction to this chapter, this recipe is not appropriate for Tabs 1 through 5. It combines a very pale amber base color with a highlight. It's girlish and feminine—innocent in a not-so-innocent way! This recipe uses the TTR technique, which I believe changes at-home hair-color application, bringing it to a near-professional result level! Natural blondes can use the Lemon Blossom recipe for highlights, without the base color.

Tabs 5½ and 6

MATERIALS

Vaseline petroleum jelly

1 box light golden blonde

Timer

Shampoo and conditioner

Standard highlighting recipe (Chapter 3, page 44)

Cotton balls

Oral-B toothbrush, soft, with an elongated tip

Tung brush

Plastic wrap

Hair dryer with diffuser or diffuser sock

Clean washcloth

Deep olive oil conditioner (page 279)

DIRECTIONS

1. Detangle your hair and comb it away from your face.

2. Prepare the light golden blonde tint according to the package instructions.

3. Apply approximately ¾ the mixture to your hair starting 1½ inches down from your scalp through to the ends. Comb through your hair to make sure the mixture is distributed evenly.

4. Set the timer for 5 minutes.

5. When time is up, quickly apply the remainder of the mixture starting from your scalp to 1½ inches down.

6. Set the timer for 2 minutes.

7. When time is up, rinse your hair, shampoo gently with no scalp manipulation, condition, and dry thoroughly.

8. Mix up the standard highlighting recipe (Chapter 3, page 44).

9. Part your hair the way you like best.

10. If you wear your hair off your face, the application begins ½ inch from your ear all the way around to your other ear. If you wear your hair down with a fringe or bangs, start in the middle of the bangs where the bang connects to the regular hair.

11. Apply the highlights with the TTR technique (thread, thread, ribbon) found on page 48: Pick up a ½-inch ribbon of hair closest to the right side of your head. Hold it at its midsection. Dip the Oral-B toothbrush into the highlighting mixture and slide it on either side of the ribbon as if you were drawing a border.

12. Put down the Oral-B toothbrush and pick up your Tung brush. Dip it into the same highlighting mixture and start pulling the product through the ribbon ¼ inch down from where the borders you made end. Continue pulling it down the center of the ribbon without adding any new product. Lift the Tung brush up every ¼ inch and overlap the product at the end slightly. Feather it down that way until you reach the end of the ribbon. The product will be heavier at the start and very light at the ends—opposite of the other highlighting techniques, which are bottom heavy.

13. Continue highlighting ribbons this way about ½ inch apart all the way around your head until you reach the other side.

14. Make a custom plastic heat cap with two sections of 14-inch-wide plastic wrap. Lay one piece over your hair front to back and then lay the second piece side to side over your hair. Plastic sticks to itself very well, so secure the "cap" around your head and it will stay put. The plastic retains the natural heat from your head and keeps the product from drying out too much. It also forms a barrier between your hair and hair dryer with a diffuser or diffuser sock.

15. Set the timer for 7 minutes.

16. Hover over your highlights with the hair dryer and diffuser or diffuser sock, focusing the heat on each section for about 20 seconds.

17. Wait 7 minutes, then check the highlights. The color of the hair with the highlight cream should match the color of a lemon blossom, which is pale white with a tinge of creamy yellow. If you need more time, diffuse for 2 minutes, but no more.

18. Immediately remove the plastic cap and wipe off the Vaseline with a clean washcloth.

19. Rinse your hair and shampoo gently. Towel blot and then apply a deep olive oil conditioner (page 279) to your hair for about 20 minutes. Shampoo and style as usual.

NOTE: The first time you do this recipe, like any good chef, you should record what you did as well as the result you achieved so the next time you can play a little bit with the time and the application. It takes three times to make the color perfect—even at the salon, it takes me three times to get it exactly right!

Tab 6½

If you are already a beautiful reddish brown, why go blonde? You are one of the lucky few whose natural color is perfect for highlights. To add the feeling of blondeness to your hair, or if it's summertime and you want to lighten up a bit, I suggest doing *just* the TTR highlighting technique (steps 9 through 18). If you want to add more color variation on top of this, try this simple procedure.

Vaseline petroleum jelly

1 box medium golden blonde

Timer

Shampoo and conditioner

Oral-B toothbrush

Tung brush

Plastic wrap

Hair dryer with diffuser or diffuser sock

Clean washcloth

Deep olive oil conditioner (page 279)

DIRECTIONS

1. Detangle your hair and comb it away from your face.

2. Apply the Vaseline to your hairline.

3. Prepare the medium golden blonde tint according to the package instructions.

4. Apply approximately ¾ the mixture to your hair starting 1½ inches down from your scalp through to your ends.

5. Set the timer for 2 minutes.

6. When time is up, apply the remainder of the mixture starting from your scalp down 1½ inches.

7. Set the timer for 2 minutes.

8. When time is up, rinse your hair, shampoo gently with no scalp manipulation, and proceed with the highlighting as follows.

9. Apply Vaseline along your hairline, following the instructions in the standard highlighting recipe (page 44) based on whether or not you have bangs.

10. Mix up the standard highlighting recipe (see page 44).

11. Part your hair the way you like it best.

12. If you wear your hair off your face, the application begins ½ inch from your ear all the way around to your other ear. If you wear your hair down with a fringe or bangs, start in the middle of the bangs where the bang connects to the regular hair.

13. Apply the highlights with the TTR technique (thread, thread, ribbon) found on page 48: pick up a ½-inch ribbon of hair closest to the right side of your head. Hold it at its midsection. Dip the Oral-B toothbrush into the highlighting mixture and slide it on either side of the ribbon as if you were drawing a border.

14. Put down the Oral-B toothbrush and pick up your Tung brush. Dip it into the same highlighting mixture and start pulling the product through the ribbon ¼ inch down from where the borders you made end. Continue pulling it down the center of the ribbon without adding any new product. Lift the Tung brush every ¼ inch and overlap the product at the end slightly. Feather it down that way until you reach the end of the ribbon. The product will be heavier at the start and very light at the ends—opposite of the other highlighting techniques, which are bottom heavy.

15. Continue highlighting ribbons this way about ½ inch apart all the way around your head until you reach the other side.

16. Make a custom plastic heat cap with two sections of 14-inch-wide plastic wrap. Lay one piece over your hair front to back and then lay the second piece side to side over your hair. Plastic sticks to itself very well, so secure the "cap" around your head and it will stay put. The plastic retains the natural heat from your head and keeps the product from drying out too much. It also forms a barrier between your hair and hair dryer with a diffuser or diffuser sock.

17. Set the timer for 7 minutes.

18. Hover over your highlights with the hair dryer and diffuser or diffuser sock, focusing the heat on each section for about 20 seconds.

19. After 7 minutes, check the highlights. The color of the hair with the highlight cream should match the color of a lemon blossom, which is pale white with a tinge of creamy yellow. If you need more time, diffuse for 2 minutes, but no more.

20. Immediately remove the plastic cap and wipe off the Vaseline with a clean washcloth.

21. Rinse your hair and shampoo gently. Towel blot and apply a deep olive oil conditioner (see page 279) for 20 minutes. Shampoo and style as usual.

NOTE: This will give you a nice little sunny look for the summer, but with a bit of warmth that is pretty, subtle, and stays true to your natural color. Do *not* leave the color of the highlighting mixtures on for more that the stated time—and apply it as quickly as you can and rinse it out just as fast. It will be gorgeous!

BARRIER CREAM—A REMINDER

Blonde colors do not stain as much as black and brunette tints, but they can still stain, especially if you have a very pale complexion and/or dry skin. I recommend using a simple barrier of thick moisturizing cream or Vaseline petroleum jelly. Apply it with your fingertips around your face at your hairline.

Tabs 7 and 7½

MATERIALS

Vaseline petroleum jelly

1 box light reddish blonde

Measuring cups and spoons

1 box light golden blonde

Timer

Clean washcloth

Shampoo and conditioner

Standard highlighting recipe (Chapter 3, page 44)

Cotton balls

Oral-B toothbrush, soft, with an elongated tip

Tung brush

Plastic wrap

Hair dryer with diffuser or diffuser sock

L'Oréal Paris Iced Coffee Color Pulse Mousse

DIRECTIONS

1. Detangle your hair and comb it away from your face.

2. Take the developer out of the light reddish blonde kit, snip off the applicator tip, and remove the cap.

3. Using measuring cups or spoons, add 1¾ ounces (three ¼ + ⅛ color container) of the light reddish blonde tint to the developer.

4. Add ¼ ounce (⅛ color container) of the light golden blonde tint to the developer.

5. Shake the bottle to mix the materials, holding your gloved finger over the open tip.

6. Apply approximately ¾ the mixture to your hair starting 1½ inches down from your scalp through to the ends. Comb through your hair to make sure the mixture is distributed evenly.

7. Set the timer for 3 minutes.

8. When time is up, quickly apply the remainder of the mixture starting from your scalp to 1½ inches down.

9. Set the timer for 10 minutes.

10. When time is up, rinse your hair, shampoo, and apply the Iced Coffee Mousse. Set the timer for 7 minutes.

11. Apply Vaseline along your hairline, following the instructions in the standard highlighting recipe (page 44) based on whether or not you have bangs.

12. Mix up the standard highlighting recipe (see page 44).

13. Part your hair the way you like it best.

14. If you wear your hair off your face, the application begins ½ inch from your ear all the way around to your other ear. If you wear your hair down with a fringe or bangs, start in the middle of the bangs where the bang connects to the regular hair.

15. Apply the highlights with the TTR technique (thread, thread, ribbon) found on page 48: pick up a ½-inch ribbon of hair closest to the right side of your head. Hold it at its midsection. Dip the Oral-B toothbrush into the highlighting mixture and slide it on either side of the ribbon as if you were drawing a border.

16. Put down the Oral-B toothbrush and pick up your Tung brush. Dip it into the same highlighting mixture and start pulling the product through the ribbon ¼ inch down from where the borders you made end. Continue pulling it down the center of the ribbon without adding any new product. Lift the Tung brush every ¼ inch and overlap the product at the end slightly. Feather it down that way until you reach the end of the ribbon. The product will be heavier at the start and very light at the ends—opposite of the other highlighting techniques, which are bottom heavy.

17. Continue highlighting ribbons this way about ½ inch apart all the way around your head until you reach the other side.

18. Make a custom plastic heat cap with two sections of 14-inch-wide plastic wrap. Lay one piece over your hair front to back and then lay the second piece side to side over your hair. Plastic sticks to itself very well, so secure the "cap" around your head and it will stay put. The plastic retains the natural heat from your head and keeps the product from drying out too much. It also forms a barrier between your hair and hair dryer with a diffuser or diffuser sock.

19. Set the timer for 7 minutes.

20. Hover over your highlights with the hair dryer and diffuser or diffuser sock, focusing the heat on each section for about 20 seconds.

21. When time is up, check the highlights. The color of the hair with the highlight cream should match the color of a lemon blossom, which is pale white with a tinge of creamy yellow. If you need more time, diffuse for 2 minutes, but no more.

22. Immediately remove the plastic cap and wipe off the Vaseline with a clean washcloth.

23. Rinse your hair and shampoo gently. Towel blot and apply a deep olive oil conditioner (see page 279) for 20 minutes. Shampoo and style as usual.

Tab 8

MATERIALS

Vaseline petroleum jelly

1 box light reddish blonde

Timer

Clean washcloth

Shampoo and conditioner

Standard highlighting recipe (Chapter 3, page 44)

Cotton balls

Oral-B toothbrush, soft, with an elongated tip

Tung brush

Plastic wrap

Hair dryer with diffuser or diffuser sock

L'Oréal Paris Iced Coffee Color Pulse Mousse

1. Detangle your hair and comb it away from your face.

2. Apply a barrier of Vaseline around your hairline.

3. Prepare the light reddish blonde color according to the package instructions.

4. Apply approximately ¾ the mixture to your hair starting 1½ inches down from your scalp through to the ends. Comb through your hair to make sure the mixture is distributed evenly.

5. Set the timer for 3 minutes.

6. When time is up, quickly apply the remainder of the mixture starting from your scalp to 1½ inches down.

7. Set the timer for 10 minutes.

8. When time is up, remove the Vaseline with a clean washcloth, rinse your hair, and shampoo thoroughly.

9. Apply the Iced Coffee Mousse according to the package directions.

10. Set the timer for 7 minutes.

11. When time is up, shampoo, condition, and dry your hair thoroughly.

12. Apply Vaseline along your hairline, following the instructions in the standard highlighting recipe (page 44) based on whether or not you have bangs.

13. Mix up the standard highlighting recipe (see page 44).

14. Part your hair the way you like it best.

15. If you wear your hair off your face, the application begins ½ inch from your ear all the way around to your other ear. If you wear your hair down with a fringe or bangs, start in the middle of the bangs where the bang connects to the regular hair.

16. Apply the highlights with the TTR technique (thread, thread, ribbon) found on page 48: pick up a ½-inch ribbon of hair closest to the right side of your head. Hold it at its midsection. Dip the Oral-B toothbrush into the highlighting mixture and slide it on either side of the ribbon as if you were drawing a border.

17. Put down the Oral-B toothbrush and pick up your Tung brush. Dip it into the same highlighting mixture and start pulling the product through the ribbon ¼ inch down from where the borders you made end. Continue pulling it down the center of the ribbon without adding any new product. Lift the Tung brush every ¼ inch and overlap the product at the end slightly. Feather it down that way until you reach the end of the ribbon. The product will be heavier at the start and very light at the ends—opposite of the other highlighting techniques, which are bottom heavy.

18. Continue highlighting ribbons this way about ½ inch apart all the way around your head until you reach the other side.

19. Make a custom plastic heat cap with two sections of 14-inch-wide plastic wrap. Lay one piece over your hair front to back and then lay the second piece side to side over your hair. Plastic sticks to itself very well, so secure the "cap" around your head and it will stay put. The plastic retains the natural heat from your head and keeps the product from drying out too much. It also forms a barrier between your hair and hair dryer with a diffuser or diffuser sock.

20. Set the timer for 5 minutes.

21. Hover over your highlights with the hair dryer and diffuser or diffuser sock, focusing the heat on each section for about 20 seconds.

22. When time is up, check the highlights. The color of the hair with the highlight cream should match the color of a lemon blossom, which is pale white with a tinge of creamy yellow. If you need more time, diffuse for 2 minutes, but absolutely no more.

23. Remove the plastic cap right away and wipe off the Vaseline with a clean washcloth.

24. Immediately rinse your hair and shampoo gently. Towel blot and apply a deep olive oil conditioner (see page 279) for 20 minutes. Shampoo and style as usual.

Tab 9

Revert to your pre-gray natural hair-color tab recipe.

Tab 10

If you are closer to 35 percent gray, revert to your pre-gray natural hair-color tab recipe. If you are closer to 65 percent gray, use the recipe for Tab 11.

Tab 11

This is an excellent color treatment for the grayer tabs, and it will get as close to a salon result as possible at home.

MATERIALS

1 perm kit (true Tab 11 only)

Vaseline petroleum jelly

1 box light reddish blonde (I like Clairol Nice'n Easy)

Measuring cups and spoons

1 box light golden brown

Timer

L'Oréal Paris Iced Coffee Color Pulse Mousse

Clean washcloth

Shampoo and conditioner

Standard highlighting recipe (Chapter 3, page 44)

Cotton balls

Oral-B toothbrush, soft, with an elongated tip

Tung brush

Plastic wrap

Hair dryer with diffuser or diffuser sock

DIRECTIONS

1. Tab 11 must pre-soften (true Tab 10s, proceed to step 2 without pre-softening; if you are a Tab 10 leaning closer to a 9 or 11, stay here): Apply the perm-kit solution to your hair starting about 1¼ inches from your scalp down to the ends and let it work for 1 minute. Rinse, shampoo, towel blot and comb out your hair. You are ready to continue with your recipe.

2. Apply the Vaseline barrier around your hairline.

3. Take the developer out of the light reddish blonde kit, snip off the applicator tip, and remove the cap.

4. Using measuring cups or spoons, add 1 ounce (½ color container) of the light reddish blonde tint to the developer.

5. Add 1 ounce (½ color container) of the light golden brown tint to the developer.

6. Shake the bottle to mix the materials, holding your gloved finger over the open tip.

7. Apply the mixture to your hair starting at your scalp down through the ends. Comb through to ensure even coverage.

8. Set the timer for 30 minutes.

9. When time is up, rinse your hair and shampoo.

10. Apply the Iced Coffee Mousse according to the package directions.

11. Set the timer for 3 minutes.

12. When time is up, shampoo, condition, and dry your hair thoroughly.

13. Apply Vaseline along your hairline, following the instructions in the standard highlighting recipe (page 44) based on whether or not you have bangs.

14. Mix up the standard highlighting recipe.

15. Part your hair the way you like it best.

16. If you wear your hair off your face, the application begins ½ inch from your ear all the way around to your other ear. If you wear your hair down with a fringe or bangs, start in the middle of the bangs where the bang connects to the regular hair.

17. Apply the highlights with the TTR technique (thread, thread, ribbon) found on page 48: pick up a ½-inch ribbon of hair closest to the right side of your head. Hold it at its midsection. Dip the Oral-B toothbrush into the highlighting mixture and slide it on either side of the ribbon as if you were drawing a border.

18. Put down the Oral-B toothbrush and pick up your Tung brush. Dip it into the same highlighting mixture and start pulling the product through the ribbon ¼ inch down from where the borders you made end. Continue pulling it down the center of the ribbon without adding any new product. Lift the Tung brush every ¼ inch and overlap the product at the end slightly. Feather it down that way until you reach the end of the ribbon. The product will be heavier at the start and very light at the ends—opposite of the other highlighting techniques, which are bottom heavy.

19. Continue highlighting ribbons this way about ½ inch apart all the way around your head until you reach the other side.

20. Make a custom plastic heat cap with two sections of 14-inch-wide plastic wrap. Lay one piece over your hair front to back and then lay the second piece side to side over your hair. Plastic sticks to itself very well, so secure the "cap" around your head and it will stay put. The plastic retains the natural heat from your head and keeps the product from drying out too much. It also forms a barrier between your hair and hair dryer with a diffuser or diffuser sock.

21. Set the timer for 7 minutes.

22. Hover over your highlights with the hair dryer and diffuser or diffuser sock, focusing the heat on each section for about 20 seconds.

23. After 7 minutes and check the highlights. The color of the hair with the highlight cream should match the color of a lemon blossom, which is pale white with a tinge of creamy yellow. If you need more time, diffuse for 2 minutes, but no more.

24. Immediately remove the plastic cap and wipe off the Vaseline with a clean washcloth.

25. Rinse your hair and shampoo gently. Towel blot and apply a deep olive oil conditioner (see page 279) for 20 minutes. Shampoo and style as usual.

NOTE: If your hair is very resistant and you are not getting the coverage you want, next time add ½ ounce of the light golden brown tint and do not add the ½ ounce of the dark golden blonde tint. If it's still not quite right, the third time you color add 1 ounce of light reddish blonde to 1 ounce of the light golden brown tint and ½ teaspoon of the medium golden tint. That should work for the most color-resistant Tabs 9 and 10. Then proceed to your highlights.

VANILLA DREAMSICLE CREAMSICLE

This is a very pale highlight. It uses a similar technique and formula described for Lemon Blossom Ribbons, but it involves the addition of 40-volume peroxide to the developer and longer work time. It's a very dramatic highlight and not appropriate or even achievable for Tabs 1 through 5½ or Tab 6½.

Tab 6

MATERIALS

Cotton balls

Vaseline petroleum jelly

40-volume peroxide developer

Standard highlighting recipe (Chapter 3, page 44)

Timer

Oral-B toothbrush, soft, with an elongated tip

Tung brush

Plastic wrap

Hair dryer with diffuser or diffuser sock

Clean washcloth

Shampoo and conditioner

1 box dark blonde

Measuring cups and spoons

Distilled water

Clairol Shimmer Lights Shampoo

DIRECTIONS

1. Make sure your hair is completely dry and parted in the style you like best.

2. Apply Vaseline along your hairline, following the instructions in the standard highlighting recipe (page 44) based on whether or not you have bangs.

3. Mix up the standard highlight recipe (Chapter 3, page 44), but using the cap of the 40-volume peroxide developer as a measure, remove 2 capfuls of the peroxide developer provided in the highlighting kit and add 2 cheater caps of the 40-volume peroxide developer to the kit developer. This provides the dream and the cream! With the kit cap securely screwed on to the kit developer, shake the bottle to combine the materials.

4. If you wear your hair off your face, the application begins ½ inch from your ear all the way around to the other ear. If you wear your hair down with a fringe or bangs, start in the middle of the bangs where the bangs connect to the regular hair.

5. Apply the highlight with the TTR technique (thread, thread, and ribbon) on page 48: Pick up ½-inch ribbon of hair closest to the right side of your head. Hold it at its midsection. Dip the Oral-B toothbrush into the highlighting mixture and slide it on either side of the ribbon as if you were drawing a border from the midsection down to the ends.

6. Put down the Oral-B toothbrush and pick up your Tung brush. Dip it into the same highlighting mixture and start pulling the product through the ribbon ¼ inch down from where the borders you made end. Continue pulling it down the center of the ribbon without adding any new product. Slightly lift the Tung brush up every ¼ inch and overlap the product at the end. Feather it down that way until you reach the end of the ribbon. The product will be heavier at the start and very light at the end—opposite of the other highlighting techniques, which are bottom heavy.

7. Continue highlighting ribbons this way all the way around your head at about 1½-inch intervals.

8. Make a custom plastic heat cap with two sections of 14-inch-wide plastic wrap. Lay one piece over your hair from the front to the back and then lay the second piece side to side over your hair. Plastic sticks to itself very well, so just secure the "cap" around your head, and it will stay put. The plastic retains the natural heat from your head and keeps the product from drying out too much. It also forms a barrier between your hair and the dryer with diffuser or diffuser sock.

9. Set the timer for 10 minutes.

10. Hover over your highlights with the hair dryer and diffuser or diffuser sock, focusing the heat on each section for about 20 minutes.

11. When time is up, check the highlight. The color of your hair with the highlight cream should match the color of a lemon blossom, which is a very pale white with a tinge of creamy yellow. If you need more time, diffuse dry your hair for 2 minutes to help kick up the highlight and then check again. If the highlight is still not where you want it, you can leave the mixture on for 25 to 30 minutes, but continue to check your color every 5 minutes. The highlight should look like cream.

12. When you have achieved the desired color, immediately remove the plastic cap and the cotton (if applicable) and wipe off the Vaseline with a clean washcloth.

13. Rinse your hair, shampoo gently, and detangle.

14. While your hair is still damp, prepare the dark blonde color by using measuring cups and spoons to remove 1 ounce of the developer and then add 1 ounce of distilled water to the developer. Add the dark blonde tint to the developer and shake the bottle to combine the materials. Apply approximately ¾ the mixture to your hair starting from your scalp to about 1 inch and wait 1 minute. Then pull the rest of the mixture through to the ends as quickly as possible with your gloved hands. Leave it in for no more than 1 minute.

15. When time is up, rinse your hair and shampoo gently with no scalp manipulation. Condition lightly and style as usual.

16. If you want, you can use a dime-size dollop of Clairol Shimmer Lights Shampoo every third shampoo.

NOTE: The first time you do this recipe, like any good chef you should record what you have done as well as the result you have achieved. Next time you can play a little bit with the time and the application. It takes three applications to make the color perfect—even at the salon, it takes me three times to get it exactly right!

Tabs 7, 7½, and 8

Cotton balls

Vaseline petroleum jelly

Standard highlighting recipe (Chapter 3, page 44)

40-volume peroxide developer

Timer

Oral-B toothbrush, soft, with an elongated tip

Tung brush

Plastic wrap

Hair dryer with diffuser or diffuser sock

Clean washcloth

Shampoo and conditioner

1 box dark blonde

Children's medicine dropper

Measuring cups and spoons

Distilled water

Clairol Shimmer Lights Shampoo

DIRECTIONS

1. Make sure your hair is completely dry and parted in the style you like best.

2. Apply Vaseline along your hairline, following the instructions in the standard highlighting recipe (page 44) based on whether or not you have bangs.

3. Mix up the standard highlight recipe (page 44), but using the cap of the 40-volume peroxide developer as a measure, remove 2 capfuls of the peroxide developer provided in the highlighting kit and add 2 cheater caps of the 40-volume peroxide developer to the kit developer. This provides the dream and the cream! With the kit cap securely screwed on to the kit developer, shake the bottle to combine the materials.

4. If you wear your hair off your face, the application begins ½ inch from your ear all the way around to the other ear. If you wear your hair down with a fringe or bangs, start in the middle of the bangs where the bangs connect to the regular hair.

5. Apply the highlight with the TTR technique (thread, thread, ribbon) on page 48: Pick up ½ inch of ribbon of hair closest to the right side of your head. Hold it at its midsection. Dip the Oral-B toothbrush into the highlighting mixture and slide it on either side of the ribbon as if you were drawing a border from the midsection down to the ends.

6. Put down the Oral-B toothbrush and pick up your Tung brush. Dip it into the same highlighting mixture and start pulling the product through the ribbon ¼ inch down from where the borders you made end. Continue pulling it down the center of the ribbon without adding any new product. Slightly lift the Tung brush up every ¼ inch and overlap the product at the end. Feather it down that way until you reach the end of the ribbon. The product will be heavier at the start and very light at the end—opposite of the other highlighting techniques, which are bottom heavy.

7. Continue highlighting ribbons this way all around your head at about 1½-inch intervals.

8. Make a custom plastic heat cap with two sections of 14-inch-wide plastic wrap. Lay one piece over your hair front to back and then lay the second piece side to side over your hair. Plastic sticks to itself very well, so secure the "cap" around your head and it will stay put. The plastic retains the natural heat from your head and keeps the product from drying out too much. It also forms a barrier between your hair and hair dryer with a diffuser or diffuser sock.

9. Set the timer for 7 minutes.

10. Hover over your highlights with the hair dryer and diffuser or diffuser sock, focusing the heat on each section for about 20 seconds.

11. When time is up, check the highlight. The color of the hair with the highlight cream should match the color of a lemon blossom, which is pale white with a tinge of creamy yellow. If you need more time, add 5 minutes more, using your hair dryer and diffuser for 2 of the minutes. If the highlight is still not where you want it, you can leave the mixture on for 20 minutes total, but continue to check your color every 3 minutes.

12. When you have achieved the desired color, immediately remove the plastic cap and the cotton and wipe off the Vaseline with a clean washcloth.

13. Rinse your hair, shampoo gently, and detangle.

14. While your hair is still damp, prepare the dark blonde color by using the children's medicine dropper to remove 1 ounce of the developer and then add 1 ounce of distilled water to the developer. Add approximately ¾ the dark blonde tint to the developer and shake the bottle to combine the materials. Apply the mixture to your hair starting from your scalp to about 1 inch and wait 1 minute. Then pull the rest of the mixture through to the ends as quickly as possible with your gloved hands. Leave it in for no more than 1 minute.

15. When time is up, rinse your hair and shampoo gently with no scalp manipulation. Condition lightly and style as usual.

16. If you want, you can use a dime-size dollop of Clairol Shimmer Lights Shampoo every third shampoo.

Tabs 9 and 10

MATERIALS

1 recipe Toasted Wheat (see page 153)

1 perm kit (Tab 9 only)

Cotton balls

Vaseline petroleum jelly

Standard highlighting recipe (Chapter 3, page 44)

40-volume peroxide developer

Timer

Oral-B toothbrush, soft, with an elongated tip

Tung brush

Plastic wrap

Hair dryer with diffuser or diffuser sock

Clean washcloth

Shampoo and conditioner

1 box dark blonde

Children's medicine dropper

Measuring cups and spoons

Distilled water

Clairol Shimmer Lights Shampoo

DIRECTIONS

1. Complete the recipe for Toasted Wheat.

2. Tab 9 must pre-soften. Apply the perm-kit solution to your hair starting about 1¼ inches from your scalp down to the ends and let it work for 1 minute. Rinse, shampoo, towel blot, and comb out your hair. You are ready to continue with the recipe.

3. Make sure your hair is completely dry and parted in the style you like best.

4. Apply Vaseline along your hairline, following the instructions in the standard highlighting recipe (page 44) based on whether or not you have bangs.

5. Mix up the standard highlight recipe (page 44). Using the cap of the 40-volume peroxide developer as a measure, remove 2 capfuls of the peroxide developer provided in the highlighting kit and add 2 cheater caps of the 40-volume peroxide developer to the kit developer. This provides the dream and the cream! With the kit cap securely screwed on to the kit developer, shake the bottle to combine the materials.

6. If you wear your hair off your face, the application begins ½ inch from your ear all the way around to the other ear. If you wear your hair down with a fringe or bangs, start in the middle of the bangs where the bangs connect to the regular hair.

7. Apply the highlight with the TTR technique (thread, thread, ribbon) on page 48: Pick up ½-inch ribbon of hair closest to the right side of your head. Hold it at its midsection. Dip the Oral-B toothbrush into the highlighting mixture and slide it on either side of the ribbon as if you were drawing a border from the midsection down to the ends.

8. Put down the Oral-B toothbrush and pick up your Tung brush. Dip it into the same highlighting mixture and start pulling the product through the ribbon ¼ inch down from where the borders you made end. Continue pulling it down the center of the ribbon without adding any new product. Slightly lift the Tung brush up every ¼ inch and overlap the product at the end. Feather it down that way until you reach the end of the ribbon. The product will be heavier at the start and very light at the end—opposite of the other highlighting techniques, which are bottom heavy.

9. Continue highlighting ribbons this way all around your head at about 1½-inch intervals.

10. Make a custom plastic heat cap with two sections of 14-inch-wide plastic wrap. Lay one piece over your hair front to back and then lay the second piece side to side over your hair. Plastic sticks to itself very well, so secure the "cap" around your head and it will stay put. The plastic retains the natural heat from your head and keeps the product from drying out too much. It also forms a barrier between your hair and hair dryer with a diffuser or diffuser sock.

11. Set the timer for 7 minutes.

12. Hover over your highlights with the hair dryer and diffuser or diffuser sock, focusing the heat on each section for about 20 seconds.

13. When time is up, check the highlight. The color of the hair with the highlight cream should match the color of a lemon blossom, which is pale white with a tinge of creamy yellow. If you need more time, add 5 minutes more, using your hair dryer and diffuser for 2 of the minutes. If the highlight is still not where you want it, you can leave the mixture on for 20 minutes total, but continue to check your color every 3 minutes.

14. When you have achieved the desired color, immediately remove the plastic cap and the cotton and wipe off the Vaseline with a clean washcloth.

15. Rinse your hair, shampoo gently, and detangle.

16. While your hair is still damp, prepare the dark blonde color by using the children's medicine dropper to remove 1 ounce of the developer and then add 1 ounce of distilled water to the developer. Add the dark blonde tint to the developer and shake the bottle to combine the materials. Apply approximately ¾ the mixture to your hair starting from your scalp to about 1 inch and wait 1 minute. Then pull the rest of the mixture through to the ends as quickly as possible with your gloved hands. Leave it in for no more than 1 minute.

17. When time is up, rinse your hair and shampoo gently with no scalp manipulation. Condition lightly and style as usual.

18. If you want, you can use a dime-size dollop of Clairol Shimmer Lights Shampoo every third shampoo.

Tab 11

This highlight is not appropriate for Tab 11.

VANILLA'D BUTTERSCOTCH

Can you hear the sirens? This barely legal color is white-hot and for only the bravest of blondes. That means the result is a *very* extreme look and not appropriate or achievable for Tabs 1 through 5 or 9 through 11. Vanilla'd Butterscotch is also not suitable for very short hair or hair with a lot of short layers. It works best if hair is shoulder length or longer and of one length or with long layers.

Tabs 5½, 6, 6½, 7, 7½, and 8

NOTE: Application of this recipe should be done on dry hair that has not been shampooed for twenty-four hours. You will get the best results if you do not use this color recipe during your menstrual cycle.

MATERIALS

Vaseline petroleum jelly

Rattail comb

Butterfly clip

1 box lightest blonde

Plastic wrap

1 box Clairol Born Blonde Nice'n Easy Maxi Bleach

2–3 packets of Sweet'n Low (or 2 teaspoons of loose Sweet'n Low), depending on how sensitive your skin is

1 lemon (to match color)

Distilled water

Measuring cups and spoons

1 box Revlon ColorSilk Ultra Light Natural Blonde #04

Shampoo and conditioner

1. Place a Vaseline barrier around your hairline.

2. Using your rattail comb, pull the top layer of your hair up by drawing an imaginary line all the way around your head at temple level, starting at the left front. That's the layer of hair you want to pull up and secure above the part on top of your head with a butterfly clip (See illustration on page 231).

3. Prepare the lightest blonde tint according to the package instructions.

4. Apply the lightest blonde hair color to the lower part of your hair starting from your scalp through to the ends. Comb through to ensure even coverage.

5. Place a 14-inch section of plastic wrap on top of the hair you have just colored. It will stick to itself, so there is no need to clip it.

6. Release the hair on top of your head and let it drop on top of the plastic wrap.

7. Prepare the Maxi Bleach according to the package directions and add 2 packets of Sweet'n Low to the mixture before shaking it to blend the materials. If your skin is very sensitive, add an extra packet of Sweet'n Low. It does an excellent job of protecting sensitive scalps.

8. Apply the Maxi Bleach to your hair starting ½ inch from your scalp down through the ends.

9. Tabs 5½ and 6 should set their timers for 5 minutes. Tabs 7, 7½, and 8 should set their timers for 10 minutes.

10. When time is up, all tabs should apply the remainder of the Maxi Bleach to their hair starting from the scalp to ½ inch down.

11. Tabs 5½ and 6 should leave the entire mixture in for 15 minutes for a total of 20 minutes' work time. The color of the highlight should match the inside of a lemon rind. If you live in a hot climate, the highlights will work faster; in colder climates it

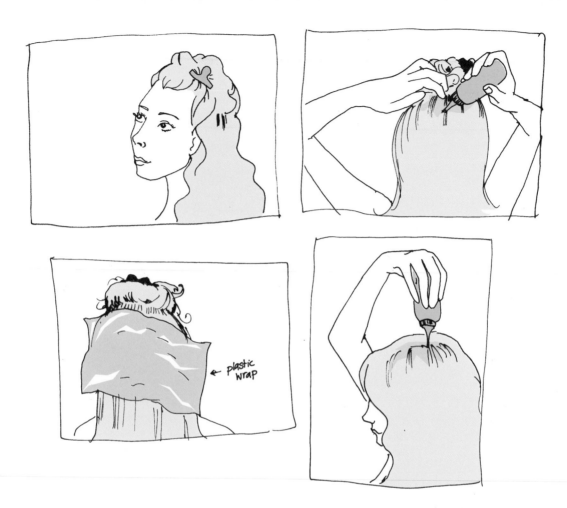

plastic wrap

may take a bit longer, which is why you have to check. If it's not at the desired color, leave it in an additional 15 minutes, checking every 5 minutes! Tabs 7, 7½, and 8 should check their results at 5-minute intervals after the initial 10 minutes, with a maximum of 30 minutes' total work time.

12. When time is up, remove the Vaseline barrier with a clean washcloth, rinse your hair, and shampoo gently with no scalp manipulation. Towel blot and detangle.

13. Prepare the Revlon ColorSilk Ultra Light Natural Blonde #04 after-gloss by using the measuring cups and spoons to remove 1 ounce of the developer and then add

1 ounce of distilled water to the developer. Add the natural blonde tint to the developer and shake the bottle to combine the materials.

14. Apply another Vaseline barrier around your hairline and apply approximately ¾ the mixture to your hair starting from your scalp down 1 inch and wait 1 minute. Then with your gloved hands, pull the rest of the mixture through the ends as quickly as possible. Leave it in for no more than 2 minutes.

15. Remove the Vaseline barrier with the washcloth. Rinse your hair and shampoo gently with no scalp manipulation. Style as usual.

BRANDIED FIG

This movie star white blonde reminds me of Jean Harlow or Gwen Stefani. It's a gorgeous beige color with a subtle hint of the palest part of a variegated fig. It's perfect for naturally fair-haired angels who are both naughty and nice! This is not an at-home option for Tabs 1 through 5 and 9 through 11. If you are no more than 65 percent gray and your natural color tab is a 7, 7½, or 8, you can follow this recipe. Otherwise, find a great salon if you must, but even then, don't expect miracles. This is a white-hot blonde, and it is not for everyone!

Tabs 5½, 6, 6½, 7, 7½, and 8

MATERIALS

Vaseline petroleum jelly

1 box Clairol Born Blonde Nice'n Easy Maxi Bleach (if you have very long hair, four inches or more below your shoulders, you may need up to four boxes to get the coverage you need)

Timer

1 lemon or 1 banana to match color

1 box light golden blonde

1 box light beige blonde

Measuring cups and spoons

Distilled water

DIRECTIONS

1. Apply a Vaseline barrier around your hairline.

2. Prepare the Maxi Bleach according to the package instructions, and apply approximately ¾ the mixture to dry hair starting ½ inch down from your scalp through to the ends.

3. Set the timer for 10 minutes.

4. When time is up, apply the remaining bleach to your hair starting from your scalp to 1 inch down to meet the mixture you have already applied.

5. Set the timer for 20 minutes.

6. When time is up, check the color. If it matches the inside of a banana or the inside of a lemon rind, you are ready to remove the Vaseline barrier with a clean washcloth. Rinse your hair and shampoo gently with no scalp manipulation. Towel blot your hair and detangle. If you have not reached your desired color, keep checking every 5 minutes until the desired effect is achieved.

7. Remove the developer from the light golden blonde, snip off the applicator tip, and remove the cap.

8. Using measuring cups or spoons, remove 1 ounce (½ developer container) of the developer and replace it with 1 ounce of distilled water.

9. Add 1 ounce (½ color container) of the light golden blonde tint to the developer.

10. Add 1 ounce (½ color container) of the light beige blonde tint to the developer.

11. Shake the bottle to combine the materials, holding a gloved finger over the open tip.

12. Apply another Vaseline barrier around your hairline and apply the color starting from your scalp through to the ends and comb through your hair to make sure the mixture is distributed evenly.

13. Set the timer for 3 minutes.

14. When time is up, remove the Vaseline barrier, rinse your hair, and shampoo gently with no scalp manipulation. Condition lightly and style as usual.

9.

THE TEEN SECTION

Teens, please feel free to use *any* recipe for your tab in this book. But teenage hair is so beautiful and full of natural highlights that you may want to think twice about covering it all up with an all-over tint. Instead, what about using one of the all-over-color recipes as beautiful subtle or bold placement for "truly you" highlights? These recipes are great for adults as well. Simply mix up an all-over-color recipe as directed, and then follow the standard highlighting technique in Chapter 3 (page 44) to apply the mixture. *Voilà!* Instead of an all-over Milk Chocolate or Spiced Persimmon color, you can use the color as threads, ribbons, or panels over your natural tresses!

And here are a few fun techniques that make hair color one big party! Hey, why not grab a friend (or two or three) and make a color session into a *real* party? Just be sure to clean your workspace and toss away used bottles and other unneeded items when you're through. It's always fun to socialize, and it's usually more enjoyable to try something new with friends who are sharing ideas, offering support, and trying out some new hair color effects on themselves.

PEEKABOO PUNK

Are you ready to add a funky new hue to your hair? This paneling technique is your chance to experiment with outrageous colors like purple, blue, or green. It's all about placement with this fun technique—if you want to show off your splash of color, you can place it right near your face. Or, you can be discreet, and keep the color hidden under a panel of your natural hair, and let it peek out just a bit whenever you like. You're in the driver's seat! Have fun with it. Since this is a very bold color effect, any tab can do this— if they dare!

MATERIALS

One butterfly clip or one yoyette

Triple-barrier cream

One box of the following: L'Oréal Paris Color Ray Fuchsia Flash, Red Rays, Copper Craze, Way Out White, Blonde Blowout, or Garnier 100% Color in Blue Black 210

Heavy-duty foil

READ ME! PREPARATION IS EVERYTHING

Before you begin any recipe, read through the ingredient list, get out your color kit, and prepare your work area. Clear off the countertop on which you will be working and spread out your materials. Detangle your hair without manipulating the scalp. Comb it out. Put on your button-down work shirt and wrap a towel around your neck and secure with a butterfly clip to catch stains. Apply the triple-barrier cream around your hairline (page 37), especially for brunette recipes. Cover the floor with a drop cloth or old newspapers. (Remember, no plastic under foot!) Be sure to have your timer out and ready to set. *Wear kit-provided gloves when applying color!* You can recycle gloves by rinsing them inside and out, patting them dry, and sprinkling some baby powder or cornstarch on the inside for easy slip-on next time.

---

8 cotton balls

Aerosol hair spray

NOTE: Do two or three strand tests (Chapter 3, page 33) to figure out what kind of punk you want to make!

(Chapter 3, page 33)

DIRECTIONS

1. Section off 3 inches of your hair in front of either your left or right ear (what side you place your punk on is your preference) and clip it off. Apply the triple-barrier cream around your hairline. Put on gloves (these products can really stain!).

2. Prepare your color choice according to the package instructions.

3. Isolate the panel of hair below the clip that you want to color. Lay a piece of heavy-duty foil under the panel. Saturate the panel with color, starting from your scalp through to the ends, and secure the foil around the panel.

4. If you wish, do another panel elsewhere using the same method.

5. As soon as you are done with the application, saturate cotton balls with aerosol hair spray as needed and wipe away the triple-barrier cream in the direction of the hairline to avoid creating a harsh line. This ensures you will not stain your skin.

6. Set the timer for 30 minutes.

7. When time is up, remove the foil.

8. Rinse the panel until the water runs clear, then shampoo and condition. Towel dry and style as usual.

TRIPLE-BARRIER CREAM—A REMINDER

First, with your fingertips apply a layer of lotion, such as Nivea or Keri brand, around your face at your hairline. Follow that with a layer of petroleum jelly. Then along the outside border of the lotion/petroleum jelly strip, apply a layer of Carmex lip balm.

DIP TIPPING

This is the most fun dip you'll ever try—I guarantee it. The best part? It's not an entirely permanent situation—the mousse color will wash out after a couple of weeks. And within a month, the permanent High Dimension color naturally fades to a nice highlight. There's only one requirement: Your hair must be long enough to make two ponytails. Other than that, it's easy, fast, and any tab can do it. Don't forget to use your gloves. Staining is *not* pretty. And remember, you are head chef on this one—let your creative juices flow when choosing mousse colors!

MATERIALS

Fabric-covered rubber hair bands

Rattail comb

4 butterfly clips

1 box Revlon High Dimension in any color two shades lighter than your natural color tab (do not use a bleach color)

4-cup capacity plastic bowl

Measuring cups and spoons

Paper towels

Plastic wrap

1 container L'Oréal Paris Color Concentrated Color Pulse Mousse in your choice of color

DIRECTIONS

1. Detangle your hair.

2. Using a rattail comb to help you, create as many ponytails as you like, starting by parting your hair in the back and pulling the two halves forward. Secure the ponytails with covered rubber hair bands. Put on gloves.

3. Mix the color according to the package instructions. Pour the mixture into a plastic 4-cup capacity bowl (this will give you plenty of room to work).

4. Then, one at a time, pretend the ponytails are paintbrushes and dip the ends into the color mixture from 2 inches to 6 inches deep, depending on how long your hair is.

5. Wrap the ends in plastic wrap. If wrap does not stick well to your hair, hold it with a butterfly clip.

6. Set the timer for 10 minutes. (If you want something subtler, leave it on for 5 minutes.)

7. When time is up, remove plastic wrap from each ponytail and rinse the ends thoroughly under running water. (The kitchen sink may be your best bet.)

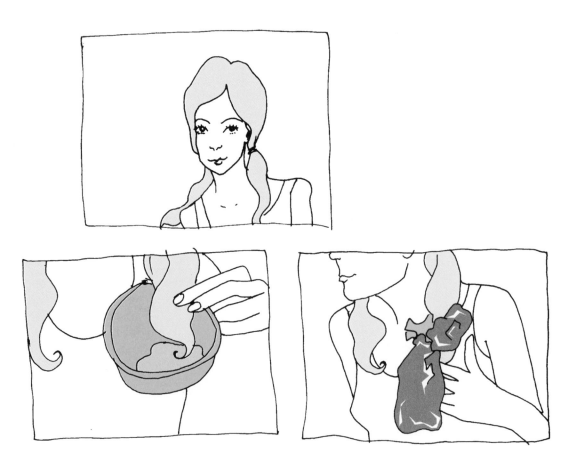

8. Keeping your hair in the ponytails, shampoo well, towel blot your ends, and detangle. Towel blot your hair again to get out any excess moisture.

9. Now put on a pair of gloves, prepare the mousse according to the package instructions, and apply it to the ends of your ponytails. Comb well.

10. Wrap the ends in plastic wrap to isolate them and to prevent staining.

11. Set the timer for 20 minutes.

12. When time is up, remove the plastic, rinse off the ends, and remove the hair bands. Rinse your hair completely, shampoo gently, and condition. Towel blot and style as usual!

NOTE: Do a strand test as described in Chapter 3 (page 33) to choose the Color Pulse mousse you like best. However, you do have some leeway for experimentation; since Color Pulse has no ammonia or peroxide; it's not permanent. Some of the great colors include Cool Blonde, Copper Blast, Iced Coffee, Hot Mahogany, and some punk chic shades such as Funky Cherry, Electric Black, Chilled Plum, and Red Pulse. I use these in the salon all the time. They rock!

THE WRAP

This is such a great technique for teens that desperately want to experiment with hair color but may not want to go too far. It's also perfect for color novices of any age! It's so subtle and easy. Because it is an uncomplicated procedure, it's also a really fun way to color hair in a group—think of a Saturday night slumber party or salon-style birthday bash. (Get out all your manicure stuff while you're at it and really have a blast!) The wrap isn't about wild, crazy color; it's just a way of adding a subtle spice to your hair to increase the grain and texture and add a little movement. You can do any color you want within two shades of your natural color, with no commitment whatsoever! It works best on at least chin-length hair of one length or on long layers.

MATERIALS

Any 1 box of color or color mousse within two shades of your own (I really like Revlon High Dimension Princess #01* on all hair color tabs.)

Heavy-duty aluminum foil cut into one to three 3-by-4-inch pieces

1–3 yoyettes

Shampoo and conditioner

*Available at beauty supply stores and online (see the Resources section on page 289)

DIRECTIONS

1. Detangle your hair.

2. Prepare the color according to the package instructions.

3. Take 1 to 3 full-length sections of hair about 2 inches in width all around your head and clip them up.

4. Unclip each piece one at a time and place a piece of foil under the section.

5. Apply the color to the hair and wrap it in foil.

6. Leave it on for the time recommended on the package.

7. Rinse your hair, shampoo gently, and condition lightly. Style as usual.

> **NOTE:** If you are "wrapping it up" with friends, use the kitchen sink sprayer and help each other with rinsing and washing. I've washed Madonna, Brad Pitt, Jennifer Aniston, and many other famous heads in their kitchen sinks! While you're waiting for the color to work, buff your fingernails or paint on some toenail polish.

10.

CLUB ROOM

MEN ONLY

Attention men! Since Louis XIV, at no time in history has our world been more open to and aware of men's grooming. Actually, I call the idea and process of coloring men's hair "regrooming." I think it's an accurate description because color for men is really a part of good grooming, staying up-to-date, and even taking some bold fashion risks.

Yes, guys, I do understand how sensitive men can be about hair color. Some men will not even set foot in a neighborhood hair salon. I don't blame them; it's a pretty bold thing to do. Even some women don't enjoy running into friends or colleagues while coloring their hair. Many men don't want to be seen buying a hair color kit in a drugstore, either! But did you know that you could order hair color online? You'll find plenty of resources at the end of this book. Or, the woman in your life—your sister, a friend, girlfriend, maybe even your mom—may pick up the color and tools you need next time she goes to the local shopping center. There must be one lucky lady you are willing to confide in!

And for you ladies reading this on behalf of your favorite guy, maybe you can talk him into adding a little color to his life. All these techniques can be done in the privacy of your own home, behind the closed doors of the bathroom. No one will ever have to know! And it is great to look and feel your most powerful.

READ ME! PREPARATION IS EVERYTHING

Before you begin any recipe, read through the ingredient list, get out your color kit, and prepare your work area. Clear off the countertop on which you will be working and, spread out your materials. Detangle hair without manipulating the scalp. Comb it out. Put on your work shirt. Guys, do not wear your best jeans or slacks when coloring hair! Put on what you wear when you are painting a room or working on a car. Wrap a towel around your neck to catch stains. Apply the triple-barrier cream around your hairline (page 37), especially for brunette recipes. Cover the floor with a drop cloth or old newspapers. (Remember, no plastic underfoot!) Be sure to have your timer out and ready to set. *Always wear your rubber gloves when applying hair color!* You can recycle gloves by rinsing them inside and out, patting them dry, and sprinkling some baby powder or cornstarch on the inside for easy slip-on next time.

Any all-over-color in this book can be used on male hair, including the gray Tab 11s out there! The techniques that follow are fun and more about fashion and taking a chance. Why not?

FROSTED SECRET SAUCE

This easy, icing-on-the-cake technique is smart and sexy. If you work on Wall Street, it may not pass the dress code. But if you want a "just got back from vacation" vibe, this look will get you halfway there! I used this technique on handsome leading man Brad Pitt for *Ocean's Eleven*. The results speak for themselves.

This procedure treats the very ends of your hair, so it works best on very short hair. That means you have to make sure you do it right after getting your hair cut, otherwise the chop shop will snip off all your hard work.

This is a two-step process. The first part, highlighting, is the same process and materials for all tabs. The second step, adding tint, is broken into two similar recipes, one for the dark tabs (1 through 5) and one for the lighter tabs (5½ through 8). Tabs 9 through 11 should follow directions for their pre-gray natural tab color recipe. Tab 9 should pre-soften (Chapter 3, page 38).

MATERIALS

1 Clairol Frost'n Tip highlighting kit

1 bottle 40-volume peroxide developer

Plastic bowl

Measuring cups and spoons

Olive oil or jojoba oil

Small spatula

Oral-B toothbrush, soft, with an elongated end

12–15 cotton balls

Vaseline petroleum jelly

Yoyette clip

Timer

Hair dryer with diffuser or diffuser sock

1 lemon (to match color)

1 box light brown (for Tabs 1 through 5)

1 box dark blonde (for Tabs 5½ through 8)

Butterfly clips

Distilled water

Shampoo and conditioner

STEP ONE DIRECTIONS (TABS 1 THROUGH 8)

1. Remove the peroxide developer from the highlighting kit, snip off the applicator tip, and remove the cap. Use the cap from the 40-volume peroxide to measure 3 capfuls of peroxide developer from the highlight kit and discard. Rinse the cap and use it to measure 3 capfuls of 40-volume peroxide, and add to the peroxide developer. Pour into the plastic bowl.

2. Add the highlight powder to the peroxide developer as per the package instructions.

3. Add 1 tablespoon of olive or jojoba oil and mix well with the spatula.

4. Put on the rubber gloves that came with the kit. Take the spatula and scoop up approximately 1 tablespoon of the mixture and put it in your gloved hand.

5. Rub the mixture in a back and forth motion all over the outermost layer of your hair, kind of like the "wax-on–wax-off" method from *The Karate Kid.* Don't use too much pressure and keep your hand parallel but always touching the hair.

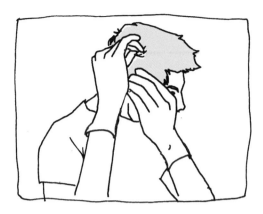 

6. Scoop up a small amount of the mixture with your thumb and forefinger and coat a few pieces around your face.

7. Use the Oral-B toothbrush to go over your hair lightly to refine and blend. Dip it into the mixture and use it to add highlights ½ inch apart along your hairline.

8. Apply Vaseline in a thin strip along your hairline. Pull and roll cotton balls into strips and stick them to the Vaseline. Take out the cap provided in the highlight kit and cover your head, securing it in the back with a yoyette clip.

9. Use the hair dryer with the attached diffuser or diffuser sock and circulate it around your head, paying more attention to your front hairline. Lighter tabs (5½ through 8) should diffuse for 10 minutes. Darker tabs (1 through 5) should diffuse for 15 minutes.

10. After diffusing, set timer for an additional 15 minutes.

11. When time is up, remove the cap and check the highlight. If it matches the inside of a fresh lemon peel, you can rinse your hair and shampoo gently with no scalp manipulation. Darker tabs may have to continue timing for up to 90 minutes to achieve the desired goal. Check every 5 minutes and be patient. Blond can take time, especially for brunettes! When you have reached your goal color, rinse hair and shampoo gently with no scalp manipulation.

12. Remove the developer from the box of light brown tint, snip off the tip, and remove the cap. Put on gloves.

13. Using measuring cups or spoons, remove 1 ounce of developer (½ developer container). Add 1 ounce of distilled water.

14. Add the light brown tint to the developer and shake the bottle to mix the materials, holding your gloved finger over the open tip.

15. Apply approximately ¾ the mixture to the nape of your neck to midcrown.

16. Set the timer for 3 minutes.

17. When time is up, apply the remaining mixture to the rest of your hair.

18. Set the timer for 2 minutes.

19. When time is up, rinse your hair, shampoo gently with no scalp manipulation, and condition lightly. Style as usual.

12. Remove the developer from the box of dark blond, snip off the tip, and remove the cap.

13. Using measuring cups or spoons, remove 1 ounce of developer (½ developer container). Add 1 ounce of distilled water.

14. Add the dark blond tint to the developer and shake it to mix the materials, holding your gloved finger over the open tip.

15. Apply approximately ¾ the mixture to the nape of your neck to midcrown.

16. Set the timer for 3 minutes.

17. When time is up, apply the remaining mixture to the rest of your hair.

18. Set the timer for 2 minutes.

19. When time is up, rinse your hair, shampoo gently with no scalp manipulation, and condition lightly. Style as usual.

NOTE: To keep these colors looking their best, order L'Oréal Paris Colorist Collection shampoo in White Violet for Tabs 5½ through 11 or Cocoa Bean for Tabs 1 through 5. Use it every third cleansing. If you like an ashier appearance, use it every other shampoo.

TRIPLE-BARRIER CREAM—A REMINDER

If you are working with dark tints, you should protect your skin with a triple-barrier cream. First, with your fingertips apply a layer of lotion, such as Nivea or Keri brand, around your face at your hairline. Follow that with a layer of petroleum jelly. Then along the outside border of the lotion/petroleum jelly strip, apply a layer of Carmex lip balm.

DELETE THE GRAY

All Tabs

Stroke your gray away with this simple procedure. It adds just the right youth serum to your hair color. The essence of this technique is to remove gray, or the salt, and add more color, or pepper. The key to this look is to leave a good amount of your natural color and to not completely cover your gray. After twenty-four years in the hair color business, I can confidently tell you that it doesn't look great when a man obviously colors his hair the wrong color. Just think of Donald Trump. Get the picture? In addition, the solid color, one-dimensional look you get from many men's hair color kits just doesn't look natural. I use this technique on power executives, hipsters, and rock stars alike. It just looks very *real*.

Finally, prepare your workspace well and wear protective work clothes. Delete the Gray can be a bit messy.

MATERIALS

Just For Men Mustache, Beard, and Sideburns formula* in a shade that correlates to your natural tab

4-cup capacity plastic mixing bowl

Oral-B toothbrush, soft, with elongated end

Timer

Shampoo and conditioner

DIRECTIONS

1. Mix the developer and the tint in a plastic mixing bowl according to the package instructions.

---

*I prefer the way the facial hair formula works for this technique—and you can use the same technique on your beard, mustache, and sideburns! The Oral-B gives you greater control than the little brush that comes in the box.

2. With your gloves *on*, dip the Oral-B toothbrush into the color, so it covers the bristles well, but not so it's falling off.

3. Brush the Oral-B across your hair in random, imperfect, and fine feathery sections all around your head. Do not do this in an obvious pattern or it will look artificial—but *do* keep the sections about equal in number on both sides. This technique is meant to give you more color and appear natural. Remember, you're in charge!

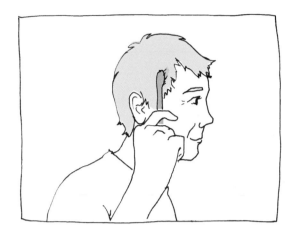

4. Set the timer for 5 minutes.

5. When time is up, rinse your hair, shampoo well, condition lightly, and style as usual.

   Voilà! Here's looking at you, kid!

NOTE: To keep these colors looking their best, buy Clairol Shimmer Lights Shampoo for Tabs 5½ through 11 or L'Oréal Paris Colorist Collection Cocoa Bean Shampoo for Tabs 1 through 5. Use it every third cleansing. If you like an ashier appearance, use it every other shampoo.

FOR MEN ONLY—*NOT!*

Ladies, if you have short hair or a stray gray here or there, these techniques work beautifully for you, too!

ROCK STAR

Any tab guy (or girl with short hair) can use this rock 'n' roll recipe—if they dare! Tab 11 may have to do the recipe twice one week apart for an optimum result. This is a *very* stark look. Be prepared for maximum stare. If you want it, it's yours! Rock Star does not pretend to be natural. It's about looking and feeling rebellious, tough, and sexy. In fact, Rock Star is the secret to the look of many of your favorite rock stars! So go ahead, tough guy . . . I dare you.

This recipe uses a standard black and the technique is messy. So you absolutely (no excuses) must make the time to apply the triple-barrier cream, put on gloves, and wear your messiest clothes. Protect your work area!

MATERIALS

Triple-barrier cream

1 box Garnier 100% in Blue Black or Nice'n Easy Blue Black

Vaseline brand petroleum jelly

8 cotton balls

Aerosol hair spray

Timer

Shampoo and conditioner

L'Oréal Paris Color Pulse Electric Black #10

DIRECTIONS

1. Detangle your hair, but go easy on the scalp.

2. Apply the triple-barrier cream and put on gloves.

3. If you love a look on the edge and want more contrast, create a Vaseline "resist": apply Vaseline with your fingers to a 1-inch chunk on your temples or off the center from your part or wherever you do not want the black color to show up. The Vaseline will act as a barrier between your hair and the dye. (Remember, this is not a technique that shows up in the boardroom very often.)

4. Prepare standard black according to package directions.

5. Apply the color mixture to your hair starting from your scalp through to the ends, avoiding the areas where you placed the Vaseline.

6. As soon as you are done with the application, saturate cotton balls with aerosol hair spray as needed and wipe away the triple-barrier cream in the direction of the hairline to avoid creating a harsh line. This ensures you will not stain your skin.

7. Set the timer for 35 minutes. You want as much buildup of color as possible so the weight of the pigment is obvious.

8. Shampoo well, condition, and towel dry.

9. Shake the can of Electric Black mousse and put on gloves. Apply the mousse to all the black areas of the hair and comb through.

10. Set the timer for 20 minutes.

11. When time is up, rinse your hair well until the water runs clear (remember, this is black dye—you do not want any excess in your hair). Shampoo well, condition lightly, towel blot, and style as usual.

SURF PUNK

So you think you're a surfer-rocker punk, do you? Does playing the Sex Pistols or Green Day do wonders for your attitude? Well then, proceed with no caution because you know what a daredevil you really are. The only caveat is your hair should be as least ear-length in front. If in doubt about this look, rent the movie *Lords of Dogtown* (Sony Pictures) and see Surf Punk in action on the boys.

Do not forget to get your split and ragged ends shaped (loosely trimmed) *before* embarking on this process. And make sure your hair is parted and dry before you begin this two-step process.

Clear and protect your workspace and get it (and yourself) ready for an adventure!

Tabs 1 through 8

MATERIALS

I box Revlon High Dimension Bleach Blond lightening kit or Frost'n Tip in Dramatic (original)

Plastic wrap

1 apricot, to match color (Tabs 1–3)

1 lemon, to match color (Tabs 5–8)

1 box medium brown for Tabs 1–3

1 box light brown for Tabs 4–5½

1 box dark blond for Tabs 6–8

2 teaspoons olive oil

Distilled water

Measuring cups and spoons

Clairol Shimmer Lights Shampoo (for Tabs 5½–11) or L'Oréal Paris Colorist Collection Cocoa Bean Shampoo* (for Tabs 1–5)

Available at beauty supply stores and online (see the Resources section on page 289)

1. Part and dry hair the way you like wearing it best.

2. Apply a thin line of Vaseline along your hairline at your forehead and temples. Unroll the cotton balls into long strips and stick them on top of the Vaseline.

3. Mix the bleach blond kit according to the package instructions and add 2 teaspoons of olive oil.

4. Put on your gloves, scoop up a two-finger dollop of the mixture, and coat 2 front sections of hair on either side of the part, starting from the top of your eyebrow down through to the ends. Be sure to coat the product on heavily, especially at the bottom.

5. Wrap the sections in plastic wrap.

6. Continue working this way in sections around the outer or top layer of your hair all around the head, midsection down through to the ends. Wrap each finished section in the plastic.

7. FOR TABS 1 THROUGH 3, start checking at 10 minutes and continue checking every 5 minutes up to 40 minutes. Leave the mixture on until it matches the color of a ripe apricot, then rinse and shampoo your hair gently with no scalp manipulation, condition lightly, and towel blot. Detangle your hair and move on to the next step.

FOR TABS 5 THROUGH 5½, start checking at 10 minutes and continue checking every 5 minutes up to 40 minutes. Leave the mixture on your hair until it matches the outside of a lemon peel, then rinse and shampoo your hair gently with no scalp manipulation, condition lightly and towel blot. Detangle your hair and move on to the next step.

FOR TABS 6 THROUGH 8, start checking at 5 minutes and continue checking every 5 minutes up to 30 minutes. Leave the mixture on your hair until it matches the inside of a lemon peel, then rinse and shampoo your hair gently with no scalp manipulation, condition lightly, and towel blot. Detangle your hair and move on to the next step.

8. **FOR TABS 1 THROUGH 3,** prepare the medium brown color by removing 1 ounce (½ developer container) using the measuring cups and spoons (discard it) and adding 1 ounce of distilled water to the developer. Add the tint and prepare it according to the package instructions. Apply the mixture to your bleached ends and leave it on for 3 minutes. Rinse your hair, shampoo gently, towel blot, and style as usual.

FOR TABS 4 THROUGH 5½, prepare the light brown color by removing 1 ounce (½ developer container) of the developer using the measuring cups and spoons (discard it) and adding 1 ounce of distilled water to the developer. Add the tint and prepare it according to the package instructions. Apply the mixture to your bleached ends and leave it on for 5 minutes. Rinse your hair, shampoo gently, towel blot, and style as usual.

FOR TABS 6 THROUGH 8, prepare the dark blond color by removing 1 ounce (½ developer container) of the developer with the measuring cups and spoons (discard it) and adding 1 ounce of distilled water to the developer. Add the tint and prepare it according to the package instructions. Apply the mixture to your bleached ends and leave it on for 2 minutes. Rinse your hair, shampoo gently, towel blot, and style as usual.

9. To keep these colors looking their best, use Clairol Shimmer Light Shampoo for Tabs 5½ through 11 or Colorist Collection in Cocoa Bean for Tabs 1 through 5. Use it every third cleansing, or if you like an ashier appearance, use it every other shampoo.

Tabs 10 and 11

This technique is really not recommended for Tabs 10 and 11; it requires too much maintenance. If you're determined, use the recipe that gets your hair back to your pre-gray color, and then go for it.

PART III

A Beautiful Finish

11.

THE FINE POINTS

A Q&A ON PREVENTATIVE MEASURES, LITTLE FIXES, AND DIVINE DETAILS

Sometimes a new client comes to me because they had a coloring mishap at home or even at another salon. Accidents happen, emotional decisions are made ("I'm depressed, I think I'll bleach my hair"), and a color catastrophe results. So many blunders can be corrected—but the first rule when an unwanted result occurs is not to panic! Most common color problems can be modified to something very pretty and acceptable until enough hair grows back and you can start again.

Here are answers to my clients' most common questions:

Q: CAN I APPLY ALL-OVER COLOR TO DRY HAIR?

A: Yes, but make sure it is detangled and combed away from your face.

Q: HOW CAN I REACH THE BACK OF MY HAIR MOST EFFECTIVELY?

A: Buddy up with a friend and help each other with applications. Or invest in a three-way mirror and secure it on the wall of your most frequently used work area, such as the bathroom.

Q: DOES COLOR "TAKE" BETTER ON DIRTY, UNWASHED HAIR?

A: No, but do not wash your hair for twenty-four hours before you apply all-over color. Your natural oils protect your scalp from unnecessary sensitivity. Can't help but wash it? Add two packs Sweet'n Low to any all-over color mixture.

Q: WILL A PERM AFFECT MY COLOR?

A: Yes. It actually helps the color to cover better, especially for Tabs 10 and 11. If you have permed hair, I recommend going a shade lighter than you really want to be. That way you know it will not grab too intensely. You can always go darker. Please wait one week after a perm to color. And if you have a perm and want highlights, *please* go to a salon.

Q: CAN I MIX DIFFERENT COLOR BRAND TINTS TOGETHER?

A: Yes, in a pinch. If you are at the store and they do not have one color from your recipe, grab the closest color from a different brand. But the manufacturers have specifically designed their products to work together, so if you are using more than one box, it is best to use the same brand. If you are using more of one tint than another, be sure to use that developer with that tint, since developers may vary. Someone wanting to go lighter with their hair will usually need a stronger developer than someone going darker or covering gray. In any case, be sure to do an allergy test.

Q: IF I USE THE SAME BRAND OF COLOR, CAN I MIX LIQUID AND CREAM TINTS?

A: The manufacturer would suggest not to, and I really do not believe in altering this rule! Keep creams with creams and liquids with liquids, unless you're in a pinch and have to get your color done and the store has run out. In any case, do an allergy test!

Q: MY TARGET COLOR CAME OUT A LITTLE DARKER THAN I LIKE. IT LIGHTENED UP AFTER THE FIRST FEW WASHINGS, AND I THOUGHT IT WAS GREAT! IT IS TIME TO TOUCH UP MY REGROWTH. WHAT DO I DO?

A: If the color has faded and you like it now, delete the darkest color in your recipe by ¼ ounce (⅛ color container) and add ¼ ounce (⅛ color container) of the lightest color. If your recipe has only one shade, go to the next lighter color within the shade. For example, if your goal color is Dark Chocolate and you're at Tab 1, next time delete ¼ ounce (⅛ color container) of dark golden brown and add ¼ ounce (⅛ color container) of medium golden brown. For your touch-ups, do the same lightening process; that way the new hair at the scalp isn't darker than the slightly more faded hair that had been colored previously. And if you want to go darker, use this process in reverse: delete ¼ ounce (⅛ color container) of the lighter shade and add ¼ ounce (⅛ color container) of the darker shade. You'll get your best shade by trial and error—it may take a few times to get it perfect!

Q: I JUST APPLIED A RECIPE TO MY GRAY LOCKS OR LIGHT HAIR IN AN EFFORT TO GO DARKER, AND I GOT A PHONE CALL FROM A LONG LOST FRIEND—AND I STAYED ON THE TELEPHONE TOO LONG! NOW WHAT?

A: Don't worry! It lightens up after the first two or three washings. Next time keep the timer with you—put it in the pocket of your old button-down work shirt. The beauty of color tints is that they actually have a life. The formula will stop working after 30 to 45 minutes, depending on the manufacturer. So if you have talked too long, rinse your hair as soon as possible.

Q: I AM BLEACHING MY HAIR AND JUST LOST TRACK OF TIME . . . NOW WHAT?

A: Check the color. If it is white or at any time your scalp is burning, rinse it off immediately! Next time, don't lose track of time. Use a timer with a loud buzz, or keep it next to you or in your pocket at all times. If you leave bleach on too long, it will get too light and break off. This is why all my bleaching and highlighting instructions tell you to check the hair regularly, every few minutes. The recipes in this book are designed for the at-home colorist. If you stay within your color range, you should be fine. That said, you *must* pay absolute attention with no distractions when bleaching your hair. If bleach does stay on too long, you must get to a salon for a haircut, deep conditioning, and corrective color. Don't try to handle this yourself. Never, ever, put any powdered bleach or highlight on the scalp; it is not designed for this use and has no buffering systems.

Q: I LEFT THE BLEACH OR HIGHLIGHT ON THE RIGHT AMOUNT OF TIME, AND MY HAIR IS NOT DAMAGED, BUT NOW THAT I'M DONE, I THINK IT'S TOO LIGHT. WHAT CAN I DO?

A: If its too light for your taste, use the Frosted Secret Sauce formula in chapter 10 (page 247) to balance it, depending on your desired results, or order L'Oréal Paris Colorist Collection shampoo in Sunflower and Walnut (for warmer tones) or Cocoa Bean (for cooler tones). Use a half-and-half combo of the two every other shampoo.

Q: IF I AM DOING A SMUDGE TO GO A BIT LIGHTER, DOES IT MATTER IF I LEAVE THE COLOR ON LONGER THAN 5 MINUTES?

A: Yes. The smudge technique, no matter what color you use, is designed for use for no more than 5 minutes on the scalp. Remember, we are bending the manufacturers' rules to get what we want. It is imperative to follow all directions in this recipe book.

Q: MY ENDS ARE BLONDE FROM BLEACH, MY NATURAL TAB IS DARKER, AND I WANT TO GO BACK TO MY NATURAL COLOR. WHAT DO I DO?

A: I suggest a trip to the salon. If that is absolutely not an option, you will find recipes for getting you in the right direction. Look up your target color and follow the recipe *exactly*. For example, Toasted Wheat and Deep Caramel are great recipes for bringing too-blonde hair back to a more natural state; to start, I suggest adding ribbons or threads above the ear and not on the part. Be sure to get your ends trimmed, too, and put a little Vaseline on the last ¼ to ½ inch of hair to keep the ends from getting an inky, drab look. Remember to rinse well until the water runs clear and then shampoo. If you choose to go from super-blonde to one of these darker shades, be sure to block out a couple of hours in your day!

Q: MY ENDS ARE DARKER THAN MY NATURAL TAB HAIR COLOR. IS IT POSSIBLE TO EVEN OUT THE COLOR BEFORE I CONTINUE WITH A RECIPE?

A: I truly believe the salon professional is better off taking care of this for you. But if you cannot get to a salon, use Color Oops. Heavily coat your darker ends with Color Oops, following the package directions. Then follow the target color tab recipe exactly.

Q: MY HAIR COLOR IS FINE, EXCEPT FOR THE GRAY AROUND MY HAIRLINE. WHAT CAN I DO?

A: Simply choose a recipe close to your natural hair color and prepare the amount of color tint suggested in the manufacturer's allergy test. Mix it with a Q-tip and apply. This can also be done along your part.

Q: WHAT DO I DO WHEN THE ENDS OF MY HAIR HAVE FADED? MY HAIR LOOKS UNNATURAL.

A: This is the perfect time to try the foil-out technique (page 41) with your recipe! If you are looking for coverage with no texture, use the corresponding color shampoo from your recipe twice at full strength (three times for maximum results). Apply the shampoo from the nape of your neck toward your hairline, then all the way through your ends. If you use Revlon Colorist color, don't forget to use their wonderful weekly gloss. Need more intense results? Try one of the L'Oréal Paris Color Pulse mousses. Intense results guaranteed.

Q: AS I GET OLDER, SHOULD I CHANGE MY HAIR COLOR?

A: Probably, if you and your life have changed! You want your hair color to enhance and complement your skin tone and eye color. Is now the time to go warmer? Are you ready for an adventure? Anytime you are ready to make a change, go back and do a color and lifestyle assessment. It's important—this is not a one-time step!

Q: IF MY HAIR IS 100 PERCENT WHITE, SHOULD I GO BLONDE?

A: Yes, but remember, it's all about maintenance. Do you want the upkeep? Can you handle it? If so, why not have a blonde adventure? Look at your eyes. Are they warm? Do a Deep Caramel with Vanilla Ribbons. Are they cool? Go for Toasted Wheat. I truly think you have to look at who you are. 100 percent white hair is gorgeous! So you decide. Because, my dear, we are all beautiful from the inside out. Embrace your style and your beauty!

A: This is simply not true, although the upkeep is easier if you do. There is a "rule" that says that gray-haired ladies cannot be beautifully brunette. I must dispel this silly myth! You must be the person you want to be: Hair color is all about feeling your most wonderful. A color must suit you, and it must help you feel good about yourself and the way you look. I do find Licorice or very dark colors can date older women or make them look older, unless they are a fashionista or a wild child at heart, and they will require more of your time. You can certainly do the chocolate brunettes and other deep, warm shades. Remember to have your looking-glass moment. What are your *eyes* telling you?

A: Whenever you know that gray works for you! Being anti–hair color is rebellious and easy. And, you can always change your mind and start coloring again whenever you like.

A: Once again, look into your eyes. Are they warm or cool? If you want the front to be as dark as the back, find the target color that most closely matches your back color, and use the recipe for your exact tab. If you want to take advantage of the front and have an adventure, then I suggest not matching the back. Pick your lighter fantasy target color. Using the recipe for Tab 11, apply it to the hair from the scalp to the ends and leave it on for 30 minutes (for resistant gray, up to 45 minutes). Then rinse your hair, shampoo gently with no scalp manipulation, towel blot, style as usual. Stunning!

A: Yes. After coloring, rinse your hair out completely and shampoo it lightly. Towel blot your hair very well and put some of the manufacturer's conditioner into your hand—1 teaspoon for short hair, or 2 or 3 teaspoons for longer hair. Add 2 drops of your favorite essential oil per teaspoon of conditioner. Lavender, orange, and peppermint are my favorites, and their fragrance blends well with the conditioners in the kits. Apply the conditioner from ½ inch from your scalp through to the ends and gently work it in. Leave it on for 15 minutes, then rinse well. Keep in mind that these are essential oils, and they are more powerful than you might think. Do use them sparingly, and keep the peppermint away from your scalp and eyes—it is the strongest of the three.

Q: CAN I CLEAN THE HOUSE OR COOK DINNER WHILE I AM WAITING FOR MY COLOR TO WORK?

A: If you must! Wrap cotton strips around your hairline and secure the rest of your hair in a butterfly clip. Take it from me, you could get spots everywhere. It happens to me, and it could happen to you. I have so many clients who don't want to wear a smock. Disaster! Wear your old button-down work shirt and try not to move around too much when your hair is full of tint. Murphy's Law says if you don't take the preventative measures of perching on a protected or washable chair as you wait for the color to work, "the blob" (i.e., color stain) will take over your favorite white sofa! Or, worse, a spot color could migrate onto loose strands and leave an unwanted patch of color!

Q: HOW DO I GET THE STAINS OFF OF MY CABINETS, CARPETS, AND CLOTHES?

A: Preventative measures are the key! As I've said before, lay old newspapers on floors and countertops and wear an old work shirt. Keep it in your at-home hair-color kit. But if you do have a mishap, spray clothing stains with aerosol hair spray and then immediately rinse the item with cold water. If color gets on a white-painted wall, dab some highlight bleach on the stain. Other colors may have to be repainted, unfortunately (which is why it is so important to protect your work area). Begin with a coat of stain-killing primer, such as Kilz. Now you know why salons keep a touch-up can of paint around at all times!

Q: I WEAR GLASSES. CAN I KEEP THEM ON WHILE MY TINT COLOR IS PROCESSING?

A: Do what we do at the salon to avoid tarnishing or staining the temples. Take small pieces of foil and wrap them along the length of the temples, or, if you're feeling fancy, buy TempleClean protective eyewear covers. You can order them online (check the Resources section, page 289) for about $8 for 160 of them—a lifetime supply for any at-home colorist. Or share them with your friends!

THE PERFECT SETTING

Truly beautiful brows are neatly groomed and beautifully shaped. Here are some tips for salon-perfect brows. The length of your brow should run from the inner corner of your eye to the outside corner of your eye. To discover where your arch should be, take a ruler or a pencil and hold it at your nostril and straight up past the iris of your eye. Take the ruler and hold it from your nostril to the outside corner of your eye. The end of your brow should gradually taper off here.

TOOLS

Good-quality tweezers

Brow brush or a child's toothbrush

Brow pencil or brow powder and brush

DIRECTIONS

1. Brush eyebrows into a shape with a brow brush or soft toothbrush. Brush in an upward motion. You can also spray your little brush with hair spray to hold your brows in place while you are working.

2. Tweeze any hairs that obviously lie outside of your brows' natural shape. If you have blonde brows, use the brow pencil or powder and brush to softly etch the shape in and tweeze the hairs outside the shape. Begin by tweezing from underneath the brow and pull hairs in the direction they are growing, in a sharp, swift movement. Pull one hair at a time, alternating from one brow to the other. This eliminates over-tweezing and achieves a more balanced look. Step back and check your work often!

Q: CAN I WEAR MY EYEGLASSES WHILE I APPLY MY HAIR COLOR?

A: Yes. If you can see with a magnifying mirror, take your glasses off and apply color around your hairline. Then put foil or TempleClean protective eyewear covers (see the question on page 267) on your glasses and do the rest of your application as directed. Pay special attention to the hair under the glasses—be sure to saturate it with color. Do not put the glasses on, take them off, and put them on again—you will trap your hair underneath and create spots. If you are doing highlights, use the TempleClean covers or foil exclusively. Put on

3. Use a brow pencil to fill in any gaps. It's best to go a bit lighter or the same shade when matching a brow powder or pencil color—anything too dark will look heavy and artificial. Use light, feathery strokes to draw in each hair, and then blend the color with a clean brow brush or a child's toothbrush (my favorite).

your glasses, pull your hair out from around your glasses, and highlight accordingly. Leave your glasses on until you are ready to rinse and shampoo. Otherwise, wear contact lenses during color application!

Q: THE TINT IS SITTING ON MY HAIR AND IT LOOKS PURPLE OR BLACK.

A: Don't worry; tints look much darker when they're working. Take a deep breath and wait for the timer to ring!

A: First of all, always wear gloves. I know from personal experience that stains can happen no matter what. The first step is soaking your fingertips in a fresh lemon for 20 minutes. This is a cool trick that I learned at Lanny Nails in Los Angeles. Cut the lemon in half and stick the stained nails right in the juicy lemon. You can also use straight lemon juice in a bowl. You will see a 70 percent improvement in the stain. Use an emery board to gently buff the remaining stain away. If the stain is severe, try a bit of Ajax or Comet on a sponge and gently rub the stain in a circular motion and buff with a nail buffer to bring back the smooth finish and shine. Moisturize afterward, because this product can dry out your hands!

Q: CAN I USE NAIL POLISH REMOVER TO LIGHTEN MY BROWS OR TO REMOVE HAIR COLOR STAINS FROM MY FACE?

A: Never! No! Do not even *think* about it. This is the most "dangerous" question I have ever been asked! A cotton ball soaked in aerosol hair spray is your best bet for removing stains from your face immediately after color application is complete.

DIVINE DETAILS

Q: I'M A GUY. I KNOW I CAN COLOR MY BEARD, MUSTACHE, AND SIDEBURNS, EVEN MY EYEBROWS, WITH A SPECIALLY MADE PRODUCT SUCH AS JUST FOR MEN. BUT MY CHEST HAIR IS GETTING A LITTLE GRAY. CAN I COLOR THAT, TOO?

A: Definitely! But if you have plans to go to the beach or to a summer picnic wearing a tank top, make sure you color 3 to 7 days before, in case your skin stains a bit. First make sure you use a good moisturizer, such as Nivea or Keri, all over your chest to provide a barrier. You can use your Just For Men color with an Oral-B toothbrush and apply it all over your chest hair. Leave the color mixture on for the recommended time and shampoo in the shower to completely remove it. If stains occur on your chest, one of my coworkers uses Windex spray on some cotton balls. It has the best stain removing power. Wild, huh?

Q: HOW DO I GET THE CREAM (TUBE) HAIR COLOR UNSTUCK FROM THE TEASPOON WHEN I AM MEASURING FOR MY STRAND/ALLERGY TEST?

A: Just pour your developer first and then use a Q-tip to scoop it out. (This gives a Teflon-like coating that helps the color not stick as much.)

Q: CAN I COLOR MY BROWS AND PRIVATE PARTS?

A: Although the manufacturer does not recommend it, yes, you can apply at-home hair color to eyebrows and private parts. But you must use caution! Definitely do an allergy test (described in the instructions that come with the at-home kits). Hey, if Marilyn could do it and lots of other movie stars, so can you! Follow precautions carefully! Never, *ever* dye your eyelashes!

Raise Some Eyebrows

You can match your eyebrows to your new hair color and *safely* apply color to your private areas for a completely natural look. If you take your hair color lighter and you are a natural brunette, I say dark blonde on your brows for 30 to 45 seconds is great. (Try 30 seconds first; you can always go lighter.) If you are going darker or covering gray, start with 7 minutes and build up to the desired color with 30 minutes as the maximum time. Light golden brown works best on redheads. An overly red brow can come across as comical.

Here's how to work on eyebrows.

MATERIALS

Vaseline petroleum jelly

Measuring spoons

Small clean plastic bowl

1 box of all-over color that matches your new hair color. In my testing I found that Garnier Nutrisse and L'Oréal Paris Couleur Experte work wonders. And Just For Men Mustache, Beard, and Sideburn Formula is also excellent and easily controlled.

Cornstarch

1 brow brush or clean mascara brush

1. Tweeze and shape your brows before you begin (see page 268 for tips and techniques). It's best to do this 24 hours before coloring.

2. Apply Vaseline below and above your brow.

3. Mix ½ teaspoon of the developer and ½ teaspoon of the tint in a bowl according to the package instructions and add ¼ teaspoon (or a little more) of cornstarch to thicken it. This is an extremely important step! We want the color to stay where we put it.

4. Using your brow brush (for blonde or gray going darker) or mascara brush (for brunette going lighter), carefully coat your brows with the mixture.

5. If you're a brunette and going lighter, watch a second hand on a watch or clock for 30 to 45 seconds. If you're going darker than your natural brow color, start at 7 minutes and go no longer than 30 minutes.

6. When time is up, use a dampened clean washcloth to remove the color completely. Rinse the washcloth, squeeze to remove the excess water and rinse again. Repeat this until all the hair color is removed.

NOTE: If the color is not light enough, you can repeat the procedure—just mix up a fresh batch. Discard any unused mixture.

Tinting "Down Under"

I have so many clients, celebrities included, who request that I pre-mix a recipe so that they can color their most personal part. And having gray down under has a strange and negative psychological effect on women (and men). All at-home color boxes say that the color is meant for head hair only, but my clients have tinted "down under" for decades safely and effectively using hair color products. So proceed at your own risk. Be sure to protect all delicate areas heavily, and never forget to do the twenty-four-hour patch test before embarking on this journey! For those planning a new pubic area "hairstyle," you may trim the same day, but wax or shave a few days before coloring or a few days after coloring. Natural redheads are the most sensitive, so don't skip your patch test!

MATERIALS

Vaseline petroleum jelly

Small clean plastic bowl

1 box of all-over color that matches your new hair color. In my testing I found that Garnier Nutrisse and Clairol Nice'n Easy work wonders. Or you can use the Just For Men Mustache, Beard, and Sideburn Formula in a color that matches your hair color—in this case you do not need the cornstarch.

Cornstarch

Oral-B toothbrush, soft, with an elongated end

DIRECTIONS

1. You must do a patch test. Apply the mixture according to the patch test recommended by the product you have chosen to use. Apply it in a small, inconspicuous area near where you are coloring.

2. If the patch test is successful, proceed by preparing the hairline with a heavy application of Vaseline. This is not the time to be frugal!

3. Mix up half the developer and half the tint in a bowl according to package instructions and add 1 tablespoon of cornstarch to thicken it. This is an extremely important step! If

you are using Just For Men Mustache, Beard, and Sideburn Formula, cornstarch is not necessary.

4. Using your Oral-B toothbrush, carefully coat your pubic hair with the mixture.

5. Set the timer for 30 minutes.

6. When time is up, use a dampened clean washcloth to remove the excess color completely. Remove the mixture in an upward direction, toward your belly button, not down. Rinse the washcloth, squeeze it to remove the excess water, and rinse it again. Repeat this until all the hair color is removed. At this point, you can step into the shower for a thorough rinse-off, shampoo, and a bit of the manufacturer's conditioner.

NOTE: If the color is not dark enough, wait 24 hours and use a different recipe.

12.

LUSCIOUS LOCKS... FOREVER

Let's face it: beautiful hair isn't beautiful unless it's shiny and healthy. Hair grows in accordance with your body's chemistry, balance, and well-being. What you put into your body shows in your hair—and healthy food, lots of water, exercise, and plenty of deep, restful sleep are elixirs for hair. Smoking, drinking too much alcohol, eating large quantities of junk food, and physical inactivity conspire against hair. Proper maintenance is also a key to good hair health. There are so many at-home treatments you can use on your hair to keep it in luxurious condition and maintain the color for as long as possible. And most treatments cost just pennies! Here's a compendium of my favorite hair-care techniques and treatments.

NOTE: If you are allergic to any ingredients in the hair-care recipes, do not use them on your hair!

HAIR CARE BASICS

Comb Right

Be gentle when you comb your hair or your child's hair. The best way to detangle hair when it is wet is to separate the hair into small sections and start from the bottom and work your way up with a detangle comb (a wide-toothed comb with a wide handle). Don't pull or yank hair—it will break. Don't brush wet hair, except with a paddle brush.

The Big Brush Off

So many of my clients tell me that they don't brush their hair because they think that it makes it fall out. Nothing could be further from the truth. You must brush your hair—use it or lose it! If you brush your hair when it is dry and detangled you will get the blood flowing to your scalp and stimulate the papilla. It's best to bend forward and allow hair to fall to the front. Using a natural bristle brush, brush from the nape of the neck over the head and down to the ends. Then, stand up straight and allow your hair to fall normally and brush from the underside of the hairline down to the ends. Now, brush the top layers into place with long, even strokes. Brush only when dry. Hair is weakest when it is wet, so be gentle. Even when your hair is dry, always detangle it before brushing.

Curl Girl

If you have wavy hair and do not want frizz, brush starting from your scalp through to the ends before shampooing, then condition deeply, and towel blot. Do not scrub your hair with a towel to dry it. Add a hair serum and leave it in while your hair air-dries. Banish frizz on dry hair by blotting ends with a damp paper towel, then let them air-dry.

Shampoo Smarts

When shampooing your hair on a daily basis, it is best to put the shampoo in your palm, lather it a bit, and then work it into your scalp. Massage your scalp with your fingertips during regular cleansing (but remember, no scalp manipulation during color process shampoos unless specified). Rinse your hair thoroughly. Unless your hair is very long and thick or very polluted after a day in the big city or at the beach or camping, there is no reason to repeat the shampoo procedure.

What Condition Is Your Conditioning In?

Conditioning is important after shampooing, especially if you have color-treated hair. Use a light conditioner if your hair is in good shape and a deep conditioner once a week or so. No matter how often you condition, squeeze the conditioner into your palms first and then apply it to your hair about an inch from your scalp. In other words, you want everyday conditioners to hit mostly the hair, not your scalp. The conditioner, unless it's one especially formulated for the scalp, can build up in the pores on your head and clog them. Rinse the conditioner out completely after 20 to 30 seconds or after you have finished your bath or shower.

Trimming and Haircuts

Trim long or blunt-cut hair on a regular basis, every 6 to 8 weeks, or when your hair starts losing its shape. For those of you trying to grow your hair out, changing styles from layers to a blunt cut, or growing out bangs, it is so important to keep damaged ends at bay. "Dusting" or just trimming the ends away every 2 months maintains a nice shape while you are in a style transition.

Beauty Sleep

Try a cotton sateen or satin pillowcase for sleeping so that the hair glides when you move in your sleep. It keeps tangles and frizz away!

Herb Help

Horsetail root is beneficial in strengthening the hair and nails. I recommend it to my clients who have lost their hair due to radiation treatment, have thinning hair, or are experiencing hair loss. Everyone should follow the recommended dose on the bottle (available at natural food and vitamin stores), usually 1 capsule 3 times a day. Clear it with your doctor if in doubt.

Man Trouble

Horsetail root is also a very important supplement for men (follow the recommended dosage) who work out a lot. Men secrete hormones when they are sweating, and if they don't rinse their hair well after a workout and then pop a baseball cap or other hat on, the oil-hormone-hat combo can quicken hair loss. Absolutely make sure your scalp is rinsed and lightly cleansed following a visit to the gym or ball field!

The Trouble with Oil

Witch hazel, available inexpensively in any drugstore or supermarket, is a fantastic, oily scalp astringent. Simply apply the witch hazel with a cotton ball, rinse your scalp with water, and blot it dry. Done!

No Time to Wash

Have you ever had such a busy day, you don't have time to shampoo your hair? Don't fret. A little dusting of baby powder rubbed in the hands or on a cotton ball and then patted on the scalp makes running late a little less frantic. Pat the area with a dry towel afterward, and be sure to check the back of your head in a mirror!

The Brunette Blues

Brunettes can get annoyed when gray starts sprouting up. There may be some resentment toward the silver strands, because a lot of brown-haired beauties have never had to color their hair. There's a freedom and mystique to not having to color your hair, and gray takes that liberty away. It gets worse: once a brunette touches her hair with a peroxide-based product, it makes it redder, and instantaneously the whole maintenance process is the bane of her existence!

If you are the girl who has never had to do a thing with her brunette mane and you're noticing gray, simply take your Oral-B toothbrush and Just For Men in your corresponding color and paint out the little bits and pieces around your part and face. Don't be tempted to pull out the strands of gray, because they tend to grow back curly or not as long and really it's worse. Leave them there. You're more gorgeous than ever!

Fly Away!

To keep flyaway hair under control—this is especially a problem in cold, dry climates—use any fabric softener dryer sheet. It does wonders to calm things down. Simply rub the sheet over your hair.

Mediterranean Cuisine Conditioner

A really fantastic once-a-week conditioner is 1 to 2 ounces of olive oil mixed with 3 to 6 drops of your favorite essential oil. As I have mentioned, lavender is very soothing. If you have very long hair, you may need 4 ounces of oil. Apply it starting from your scalp down through the ends and brush through vigorously on your scalp and tenderly on your hair to make sure it is distributed evenly. Leave it on for a minimum of 1 hour or even overnight. (Change your pillowcase in the morning!) To make it penetrate better, place a plastic bag or shower cap over your hair and heat it with the sun or a hair dryer for 5 minutes and let it sit the rest of the time. You can also simply massage the oil into dry hair and scalp. Jojoba and coconut oil are wonderful substitutions for olive oil.

Protect Your Hair Color "Wetsuit"

For all you serious swimmers and beachside beauties out there, here's an easy preventative measure for when you're having fun in the sun. Dampen your hair and then work through half a palm full of your favorite oil, using a wide-tooth comb. If you have long hair, secure it with a band. Olive, jojoba, and coconut are good choices, and readily available. The oil provides a great barrier to the terrors of chlorine, sun, and salt water. It's your homemade hair "wetsuit."

Tomato Juice Bath

This all-natural "bath" for your hair helps keep the green effects of chlorine at bay.

1. Shampoo your hair with a good cleansing shampoo, such as Suave or Johnson & Johnson baby shampoo.

2. Towel blot your hair.

3. Stand in the shower with your head tilted slightly back and bathe your hair in plain tomato juice. Saturate it completely. (You can also do this over the sink, but it's easier to clean up and control if you do it in the shower.)

4. Squeeze the excess out with your hands. If your hair is long, twist it up into a bun. If it's short, comb it back away from your face. Long or short, completely wrap your hair with plastic wrap, or put a plastic bag over your head and secure it above your forehead with a butterfly clip.

5. Sit in the sun for 5 minutes. If there's no sun available, use your hair dryer turned to the high heat/low-velocity setting. If you don't have a low-velocity setting, use a diffuser or an old sock over the end of the dryer for 5 minutes, and leave the wrap on for an additional 15 minutes.

6. Rinse, shampoo, and use your favorite deep conditioner.

Do this all-natural treatment once a week throughout the swimming season, or all year long if your water pipes are prone to chlorine. Use the cheapest bottled water you can find for your final rinse. It works wonders to clear your hair of damaging chlorine buildup.

Get the Green Out

If you swim a lot in public or private pools, you are adding a lot of chlorine to your hair, which can turn it green. That's not a great look, and it can be very damaging. Malibu 2000 is the ultimate hair colorists' antichlorine product, and it is available online (see the Resources section, page 289). Malibu brand also makes a well-water shampoo and conditioner for those of you who live in rural areas. If you use well water, you may have noticed your blonde hair developing a strange yellowing rust color. If you cannot get hold of Malibu 2000, a can of tomato juice worked into the hair removes chlorine. Prell and baking soda mixed together (see below) also does the trick. Tide detergent (that's right!) also gets the green out, as does Palmolive dishwashing liquid. Do not forget to condition using these remedies; detergents are very drying to the hair. I do not recommend any of these remedies for everyday washing, but once a week during swim season is not a bad idea.

Glorious, Gorgeous Gray Shampoo Cocktail

The best once-a-week treatment for gray hair is so simple and inexpensive to make at home. Simply mix up ½ cup of Prell shampoo with 1 tablespoon of baking soda. It's unbelievable! Use not more than once a week and use Clairol Shimmer Lights Shampoo (keep on for 10 minutes) and a deep condition afterward. This treatment will keep the yellowing at bay.

Nourishing Scalp Cleanser

You will be a believer once you try this easy concoction. It helps to balance your scalp, which is necessary if you spend a lot of time outside, especially in urban areas.

Small plastic bowl

3 tablespoons jojoba oil

4 drops tea tree essential oil*

½ cup water

½ cup moisturizing shampoo

1. Combine the jojoba and tea tree oil in a small plastic bowl and massage them into your dry scalp with your fingertips. I also advise doing 25 to 50 strokes on the scalp with a hair brush.

2. Leave it in for a minimum of 1 hour to overnight.

3. Mix the water and shampoo in a small plastic bowl. In the shower, pour onto your scalp, massage it through your scalp, and shampoo the rest of your hair. Rinse, towel blot, and style as usual.

NOTE: It is imperative that you don't simply let the shower water hit your scalp when it has been treated with any oil mixture, because it may have a tendency to "grab" onto your hair and scalp. That necessitates using a harsh shampoo and several washings, which defeats the purpose of an oil treatment!

Never put tea tree oil directly on the scalp. It's very intense and needs a carrier oil such as the jojoba oil recommended here.

Scalp Shooter Scrub

This is a very effective scrub for energizing the scalp and removing flaky skin. If you have any burns or abrasions on your scalp, wait until they are healed before you try this recipe.

Small plastic bowl

2 tablespoons sea salt (or raw sugar)

2 tablespoons jojoba oil

1 teaspoon agave tequila

2 teaspoons lemon juice

Timer

½ cup water

½ cup moisturizing shampoo

DIRECTIONS

1. Combine all the ingredients (except the water and shampoo) in a small plastic bowl.

2. Apply mixture to your head 2 inches from your hairline. Important: *Do not* start at your hairline, as the mixture could get in your eyes.

3. Massage in a circular motion all over your scalp and then move into your hairline area.

4. Set the timer for 10 minutes.

5. When you are ready to rinse, mix the water and shampoo in a small plastic bowl. In the shower, pour onto your scalp. Massage it through your scalp, then shampoo the rest of your hair. Rinse, shampoo a second time, condition, and towel blot your hair. Style as usual.

NOTE: It is imperative that you don't simply let the shower water hit the scalp when it has been treated with any oil mixture, because it may have a tendency to "grab" onto your hair and scalp. That necessitates using a harsh shampoo and several washings, which defeats the purpose of an oil treatment!

All-Natural Grappa Glosser

MATERIALS

Small plastic bowl or squeeze bottle

3–4 tablespoons (5–6 for longer hair) grapeseed oil, plus 2 tablespoons

2–3 drops (4–6 for longer hair) any essential oil you like (rosemary and lavender are especially nice in this)

Plastic bag or shower cap

Hair dryer

½ cup tap water

½ cup shampoo

1 drop peppermint essential oil

8 drops rosewood essential oil

5 cups bottled or distilled water

¼ cup white vinegar

DIRECTIONS

1. Combine the 3–6 tablespoons grapeseed oil and the essential oil of your choice in a small plastic bowl or squeeze bottle.

2. Apply about half the mixture to your scalp, starting 1 inch from the hairline.

3. Add the remaining 2 tablespoons grapeseed oil to the mixture.

4. Pour the mixture into the palm of your hand and pull it through your hair from top to bottom. Brush well.

5. Cover your hair with a plastic bag or shower cap and heat it for 5 minutes under the sun or with a hair dryer set at low.

6. Leave the mixture in for a minimum of 2 hours, or preferably overnight.

7. When you are ready to wash out the conditioner, mix the tap water and shampoo in the bowl. In the shower, pour the mixture onto your scalp before you wet your hair. Massage it through your scalp, then shampoo the rest of your hair. Rinse well.

8. Mix ¾ cup of the bottled or distilled water, the vinegar, and the peppermint and rosewood oils in the bowl. Pour the mixture liberally all over your hair, paying attention from the scalp to the ends. Pour the rest of the bottled or distilled water through your hair and style as usual.

> **NOTE:** When using any oil treatment overnight, be sure to cover your pillow with a clean, but not precious, pillowcase or towel. The oil will most certainly get on the pillow, and there's no sense ruining good linens.

South Seas Deep Conditioning Treatment

This rich oil treatment is fantastic for dry, damaged hair. It has the added benefit of giving you a little aromatherapy while it's working. The quantities listed make enough to treat short hair, shoulder length or shorter. For longer hair, double all ingredients.

MATERIALS

Small plastic bowl or squeeze bottle

¼ cup of coconut oil

2–4 drops of sandalwood, rosewood, or orange essential oil. I adore the combination of rosewood and orange oils; use 2 drops of each if you use them together.

Plastic bag or shower cap

Hair dryer

DIRECTIONS

1. Combine the coconut oil and essential oil(s) together in a bowl or plastic squeeze bottle.

2. Apply the mixture to your scalp, massage in, and comb through your hair.

3. Put a plastic bag or shower cap over your hair and heat it with the sun or a hair dryer set on medium for 5 minutes.

4. Leave the mixture in for at least 2 hours and preferably overnight.

Sonia's Hydrating Smoothie Hair Mask

This mask combines ingredients rich in essential fatty acids. It's an old family recipe from a client of mine, Sonia Frem, who jets in from Lebanon. It works wonders for dry, sun-damaged hair, or overprocessed hair. It feels so luxurious too—much like an expensive treatment at a deluxe spa. For shoulder length hair or shorter, halve the recipe.

INGREDIENTS

1 egg yolk

½ cup honey

2 tablespoons olive oil

The meat from 2 small to medium very ripe avocados. (You can ripen them in a brown paper bag for 2 days if the ones you have are not ripe.)

Medium plastic bowl or blender

Plastic bag or shower cap

Hair dryer

DIRECTIONS

1. Blend all the ingredients in a blender or mash with a potato masher or fork by hand in a plastic bowl.

2. Apply the mixture to your hair starting from your scalp to the ends, using ½-inch sections.

3. Cover your hair with a plastic bag or shower cap and heat it under the sun or with a hair dryer set at medium for 3 to 5 minutes.

4. Leave the mask on for about 1 to 2 hours, until fairly firm and dry.

5. Rinse your hair under the shower until all the dry product is out. Shampoo, use a detangling conditioner, and towel blot.

Sunburn Soother

If you're bald or have bald spots and get a scalp sunburn, fresh aloe vera is the way to go. Apply a dollop of it to the burn and leave it on overnight (it dries quickly). Rinse with tepid water the next morning.

If your scalp is very sunburned, see a dermatologist!

For a mild sunburn on your forehead and along your part, try the recipe below.

MATERIALS

2 cotton balls

White or apple cider vinegar

1 teaspoon lavender essential oil (see Resources, page 289)

Plastic bowl

2 tablespoons jojoba oil

Plastic bag or shower cap

½ cup water

½ cup moisturizing shampoo

DIRECTIONS

1. Dampen 1 cotton ball with the vinegar. Pat it on the sunburn. Let it dry for 2 or 3 minutes.

2. Dampen another cotton ball with ½ teaspoon of the lavender oil and pat it on the sunburn.

3. In the plastic bowl, combine the remaining ½ teaspoon of lavender oil and the jojoba oil. Work it into the rest of your hair. Keep it on overnight for the best benefit (use a plastic bag or shower cap over your hair).

4. In the plastic bowl, combine the water and shampoo. Apply to the hair and scalp and work in well. Then rinse thoroughly in the shower, shampoo, and towel blot.

resources

Most everything you need to use the recipes in this book can be found in local and chain drugstores or supermarkets. There are a few items, such as 40-volume peroxide developer and L'Oréal Paris Colorist Collection shampoos, that must be bought from a beauty supply shop or online. This list of resources includes contact information for these and other specialty products.

Most of the online resources offer all the tints, brushes, and other equipment you need as well. At-home ordering is perfect and discreet for men (or women) who simply do not want to be seen in the hair-color aisle at the neighborhood drugstore, or for those of you who live in rural areas where making a trip to the store is an event that needs a lot of pre-planning. You can sit at home or walk to the local library and order what you need right from the computer. Great hair color is available to everyone!

I also want to mention one more time that I do not work for any of the companies I have recommended throughout the book or in this resource section. I have tested a variety of products and ordered items from all the sources listed here. I name names because I have found them to be high quality and reliable for the at-home colorist. Have fun!

- **AMERICAN DISCOUNT BEAUTY SUPPLY** is a great source for all kinds of hair color and other beauty products. Specifically, I found cotton strips here, which are perfect for protecting the hairline on top of a barrier cream. It's much easier than making your own strips with cotton balls. With shipping, the total was $11.30 for 40 feet, which will last the at-home hair colorist two years at least! *www.adiscountbeauty.com*

- **COOKS.COM** has a convenient, easy to use and understand online conversion calculator to help you convert measurements. *www.cooks.com/rec/convert/*

- **DRUGSTORE.COM** is an online superstore that sells everything a brick-and-mortar store sells and more, including the Tung brush and an assortment of hair-color brands and shampoos at great prices. *www.drugstore.com*

- **FOLICA** offers many specialty hair-care and -color products, but specifically, Tab 10 and 11s can order Gray Magic here. Add two drops of Gray Magic to any color formulation and it covers the gray. The product also decreases color fading and makes color last longer. *www.folica.com*

- **HAIRCAREUSA.COM**, a specialty purveyor, sells L'Oréal Paris Colorist Collection colorist shampoos and Malibu 2000 products. Malibu 2000 products is the ultimate at-home hair colorist's professional tool for getting "chlorine green" out of hair as well as fighting the dulling effects of well water. *www.haircareusa.com*

- **IHERB** sells all kinds of herb oils and essential oils, including pure lavender oil and jojoba oil, among others. You can also find horsetail and silica tablets here. *www.iherb.com*

- **NATIONWIDECAMPUS.COM** has every L'Oréal Paris Color Pulse shade. It also offers Garnier Nutrisse, Garnier 100%, Just For Men, L'Oréal Excellence, L'Oréal Preference, Revlon ColorSilk, and many more useful items. Perfect for doing at-home hair color without ever leaving home! *www.nationwidecampus.com*

- **PARIS PRESENTS** tells you where my favorite brush kits by Miss Manicure are sold. *www.parispresents.com*

- **PAYNE'S BEAUTY SUPPLY** is a great source for 40-volume peroxide developer, my favorite Clairol Shimmer Lights Shampoo, and other hair-color tool-kit supplies. *www.paynesbeautysupply.com*

- **RACHELSSUPPLY.COM** has 1.25-ounce plastic bottles, which happen to be my favorite dispensers for oil used to remove too much highlight product. Go to the "bottles and containers" section and order #RB031, just $1.45 each. They carry many other products and dispenser sizes for deep oil conditioning.

- **SALLYS.COM** has great professional quality hair-care and styling products if you are interested in furthering your at-home colorist skills. You can also purchase Gray Magic here. Check store locations on this site, too—you never know; there may be one in your area you just haven't discovered yet. *www.sallys.com*

- **SHOPPING.YAHOO.COM** (search: hair color) has lots of colors to buy, including Revlon 01 Princess 10-Minute High Dimension for quick-lift highlights without bleach. *www.shopping.yahoo.com*

- **TEMPLECLEAN.COM** sells guards for the temples of your glasses—perfect for your at-home color kit. And they sell other great products, too. *www.templeclean.com*

- **TREASURED LOCKS** sells all kinds of high quality hair-care products for African American hair. They also offer scientific facts related to the challenges of gray hair and hair growth. *www.treasuredlocks.com*

- **WALGREENS.COM** is wonderful. It carries Color Oops, my favorite nondamaging hair-color remover, as well as Dark and Lovely Color Flash Mousse; Clairol, Revlon Colorist, and L'Oréal products; and great stuff for your at-home color kit. *www.walgreens.com*

acknowledgments

While writing this book I experienced a lot of twists and turns, and I must thank my amazing support team who helped me through the process. A huge thank you to my wonderful husband, Andrew; my dear daughter, Chelsea (you are my dream girl and make me so proud!); my sweet son, Spencer, who gives me fresh perspective; my mother, Connie Wilson Bojorquez; and Jon Hollinger, who first opened my eyes to the colors all around me. To Robert J. Clark, my Daddio, thank you for believing in me; David Mackey, for knowing; and Bobby Earhardt, for my perfect Tab 2. A big gracias to Rosidalia Castillo, for helping me disappear into my testing kitchen. Thanks to Douglas Goddard of DMD Studio Inc., who organized all the hands-on testing. Scott Bronson, *merci de maintenir des choses dans La Perspective. Meine liebes Garrett Swann, danke für das Gewürz.* To Katya Merrell, Tina St. John, Gretchen Hollinger, Keara Denton, and all my other test models, *merci des vos cheveux!!!*

A major thank you to my clients, who through the years have trusted me to translate the looks they crave into the colors that are right for them. I deeply appreciate you for your support over the last twenty-four years!

It has been an honor to be inspired by the greatest of the greats over the many years of my career specializing in hair color. I've had the pleasure to observe the best visionaries around—people who have made it possible for me to develop trademark techniques and colors. Cheers to Milika and Roxanne, by whom I was lucky enough to be taught the concept of balayage at the José Eber Salon. Milika, you changed my beach-blonde babe placement, and Roxanne, thanks for showing me the effect of threading soft highlights for subtle contrast. You both allowed me to interpret and translate those techniques and make them my own. Thank you, Steven Tapp, for teaching me the bold placement of paneling. Much thanks to Miss Tracey Cunningham. I am awed by your extraordinary hair-color talent and grateful for your friendship.

José Eber, Laurent DuFourg, Corey Powell, François Pelikan, Olivier Leroy, Christophe, Frida Aradottir, Jonathan Antin, and Sally Hershberger, thank you all for letting me add light and shadow to your shapes. Teddy Antolin, Cervando Maldonado, and Chris Turner, you refined my vision. It is great to be pushed to the limit of light: the way light hits the outer leaves on a tree, little children on the beach, or a river at the end of summer. Nature is my muse.

This project would not have happened without the diligent support of Carol Mann of the Carol Mann Agency; Christine Schwab, friend, client, and writer; the team at Collins, including publisher Mary Ellen O'Neill; editors Elizabeth Bewley and the tenacious, patient Cassie Jones; production editor Amy Vreeland; art director Richard Ljoenes; designers Kris Tobiassen and Amanda Kain; production manager Nyamekye Waliyaya; www.thesaurus .reference.com; and the talented Karen Kelly, my coauthor, whose patience and motivation have kept me focused. Dearest Karen, thank you for your flow and nonstop diligence.

Where would our colors be without my research team—Judy Nida, master of organizing, as well as Kerry Kirchman, Anival Morales, Sulaka Hilton, Tyron Brown, Chrystal Hewitt, Alexandra Thomas, Christine Buzas, and Pablo Perez. Thank you, Andrea Stellof and John Frieda, for giving me a great testing space to work in, and Charmaigne Breitengross of Rex. A big happy grin to the entire Neil George Salon team, especially owners Neil Weisberg and Amanda George, Adam Campbell, Guy Romeo, Alissa Tietgen, Harmony Polo, Melissa Brown, Rhea Lindsay, Justin Anderson, Kacey Welch, Tyle Mahoney, Christy Lash, Marnina Burnstein, Jen Bish, Kay Catato, Sydnee Fernandez, José and Ulma Martinez and Letty Martinez, and Ali Saman. Thnx4sharing.

index